OPTIMIZATION AND DECISION THEORY

OPTIMIZATION AND DECISION THEORY

Edited by:

Ivan Stanimirović

ARCLER PRESS

www.arclerpress.com

Optimization and Decision Theory

Ivan Stanimirović

Arcler Press
2010 Winston Park Drive,
2nd Floor
Oakville, ON L6H 5R7
Canada
www.arclerpress.com
Tel: 001-289-291-7705
001-905-616-2116
Fax: 001-289-291-7601
Email: orders@arclereducation.com

ISBN: 978-1-77361-068-9 (Hardcover)

DECLARATION

Some content or chapters in this book are open access copyright free published research work, which is published under Creative Commons License and are indicated with the citation. We are thankful to the publishers and authors of the content and chapters as without them this book wouldn't have been possible.

ABOUT THE EDITOR

Ivan Stanimirovic gained his PhD from University of Niš, Serbia in 2013. His work spans from multi-objective optimization methods to applications of generalized matrix inverses in areas such as image processing and computer graphics and visualisations. He is currently working as an Assistant professor at Faculty of Sciences and Mathematics at University of Niš on computing generalized matrix inverses and its applications.

TABLE OF CONTENTS

LIST OF CONTRIBUTORS

Ivan Stanimirović

João Marques
Marine and Environmental Sciences Centre (MARE), Department of Civil Engineering, University of Coimbra, Coimbra, Portugal

Maria Cunha
Marine and Environmental Sciences Centre (MARE), Department of Civil Engineering, University of Coimbra, Coimbra, Portugal

Dragan Savić
Centre for Water Systems, School of Engineering, Computing and Mathematics, University of Exeter, Exeter, UK

Orazio Giustolisi
Department of Civil Engineering and Architecture, Technical University of Bari, Bari, Italy

Siwar Ben Haj Hassine
Laboratory of Electronic and Microelectronic, Faculty of Sciences at Monastir, University of Monastir, 5000 Monastir, Tunisia

Mehdi Jemai
Laboratory of Electronic and Microelectronic, Faculty of Sciences at Monastir, University of Monastir, 5000 Monastir, Tunisia

Bouraoui Ouni
Laboratory of Electronic and Microelectronic, Faculty of Sciences at Monastir, University of Monastir, 5000 Monastir, Tunisia

Networked Objects Control & Communication Systems Laboratory, National Engineering School of Sousse, BP 264, Sousse Erriadh, 4023 Sousse, Tunisia

Amadeo Ascó
School of Computer Science, University of Nottingham, Nottingham NG8 1BB, UK

S. Morteza Mirdehghan
Department of Mathematics, College of Sciences, Shiraz University, Shiraz, Iran

Hassan Rostamzadeh
Department of Mathematics, College of Sciences, Shiraz University, Shiraz, Iran

JiyongOh
Graduate School of Convergence Science and Technology, AICT, Seoul National University, 1 Gwanak-ro, Gwanak-gu, Seoul 08826, Republic of Korea

NojunKwak
Graduate School of Convergence Science and Technology, AICT, Seoul National University, 1 Gwanak-ro, Gwanak-gu, Seoul 08826, Republic of Korea

Ziauddin Ursani
School of Mathematical and Computer Sciences, Heriot Watt University, Edinburgh EH14 4AS, UK
Route Monkey Ltd., Livingston EH54 5DW, UK

David W. Corne
School of Mathematical and Computer Sciences, Heriot Watt University, Edinburgh EH14 4AS, UK
Route Monkey Ltd., Livingston EH54 5DW, UK

PREFACE

The theory of decision deals with analyzing how a person chooses an action that, among a set of possible actions, leads to the best result given the preferences. If a person should invest or not in capital goods, what career a person is going to choose, what car is foing to buy, are very common problems that affect us in our daily life and to the which - in formal terms - is faced with the theory of decision. On the other hand, in recent years its influence in disciplines such as psychology and economics has been so great, along with applied mathematics, sociology, political science and philosophy - which have that it is very difficult to address some of these specialties without having a knowledge of theory of the decision.

Decision theory has become an indispensable working tool in disciplines as varied as economics, psychology, science politics, sociology or philosophy. Nevertheless, he remains a great stranger for many social scientists despite its great influence. In this work the basic elements of decision theory are presented first then focus on decision theory in situations of uncertainty. Thus, after explaining some classic decision criteria under uncertainty, we discuss the normative model of expected subjective utility (SEU). The limitations of this theory lead us to the more recent descriptive models that are Support of Herbert Simón's limited rationality theory as the model the adaptive decision maker or the theory of ecological rationality.

Life abounds in situations where it is necessary to make decisions. In some cases, the consequences of decisions depend only on one side, which makes the decision. For example, a programmer makes a decision in any programming language will encrypt algorithm that solves a given problem. However, the consequences of decisions often depend not only on one side but also on the interaction with the decisions taken by the other side, so that the outcome of the decision on the one hand depends on the decisions of others or other parties. It is often the case that such a situation characterized by conflicting-antagonistic interests of the participants in decision-making, ie. we say that the parties to make decisions in the conflict. In the game of chess, the result of the game depends not only moves one player more than another move and their interests are conflicting, because each side wants to win another.

This situation of uncertainty in decision-making in mathematical games is the field of operations research that deals with the analysis of these problems and finding optimal solutions, and is called game theory.

Game theory means the mathematical theory of decision-making process by the opponent (the participants, players) that are in conflict (conflict) or are involved in competitive conditions. The term game means a model of real conflict situations. The game can be added to the relevant rules, which define the rules of conduct of participants in the game, and the goal of game theory is that the exact mathematical algorithm analyzes the conflict situation and determine the reasonable behavior of the players and the course of the conflict, ie. to determine the optimal strategy for each of the participants in the game.

Game theory has the task of finding solutions in situations of competition, which is partially or completely conflicting interests are at least two opponents (ie. According to this theory among the participants in the game). The solution of the conflict is determined by the actions of all parties involved in the conflict. Game theory deals with situations that have the following characteristics: a) there must be at least two players; b) The game starts by having one or more players choose between defined alternatives; c) after the selection is associated with the first move, the result is determined by the situation that determines who makes the next selection and what are his options open; d) the rules of the game are certain rules for determining which specifies the mode of behavior of players; e) any move in the game ends the situation that determines the payout of each bonificiranog player (extra nine player is the one who makes choices and receive payments). There are many examples in different areas of life that can be observed and studied as a conflict situation. A good number of economic problems in the field of market - competitive relations contain conflicts of different interests, so they can be analyzed and solved using game theory.

In this book the solutions of multiobjective problems are considered through genetic algorithms, which consist of random searches in the search space marked by the restrictions, obtaining solutions increasingly efficient. In order to achieve this, two new methodologies are proposed, the first one (MOEGA), which elitism as an interesting concept not to lose the good results that are have achieved and obtain a Pareto border close to the real and the second (MOEGA-P) which considers the preferences of the decision maker in an interactive way, such that the decision maker can direct the search of the algorithm towards the area of its interest.

MOEGA get better solutions in the problem of the backpack comparing it with algorithms like the SPEA2 and NSGAII and MOEGA-P allows the decision-maker to obtain only a portion of the pareto according to your preferences acquiring knowledge of the problem, such that for it will be much easier for him to decide between this small group of alternatives. In addition to in the case of the backpack, MOEGA-P offers the decision maker alternative solutions that does not consider an algorithm without preferences and much closer to the Pareto frontier real, because restricting the search area with preferences, takes advantage of the cost to find more efficient solutions instead of looking at areas that are no longer are of interest to the decision maker.

In this book, the problem of multiobjective optimization is described, mentioning some classic techniques of optimization as well as some meta techniques - heuristics that are used to solve such problems. Since the interest of this research are the genetic algorithms, will be deepened in this type of methodology. The following chapter will describe the main methodologies of multiobjective genetic algorithms and will mention how the preferences of the decision maker. Taking into account the weaknesses and strengths of these methodologies, the next chapter describes an interactive methodological proposal where the decision-maker intervenes in the search process for better results and according to your preferences; But not before proposing another new methodology of algorithms genetic algorithm that obtains better results comparing it with the already existing ones. Then, the results obtained with the two methodological proposals are described along with the conclusions and some ideas of possible future work.

Editor

CHAPTER 1

INTRODUCTION

Ivan Stanimirović

Today, decision-making is a process that has gained importance as the world in which we are immersed changes rapidly by forcing the person(s) to make decisions (simple or complex) in short periods of time.

Making a decision is a process that becomes complex because many times who takes the decision does not have a single interest but multiple interests or objectives that you want to meet and usually these are in conflict with each other. Moreover, if this is systems where they make the decisions are dynamic systems, with imprecise information, in an environment of uncertainty and in some cases, decisions are taken by a group of people with different interests.

Operations research is a science that models these types of problems making use of mathematics and logic. Try to find a better solution (solution optimum) or several better solutions, such as to conduct the operations within a organization by allocating the most effective resources available in different activities (Hillier and Lieberman, 2002).

This science provides different tools to people who are going to take a decisions, which depend on the characteristics of the problems that are desired solve. Among these tools are linear programming, dynamic programming, whole programming, nonlinear programming, game theory, decision analysis, stochastic processes, simulation, among others.

In particular, the analysis of decisions studies the decision making with a degree of uncertainty. It provides decision-makers with methodologies that indicate how to take rationally when there is uncertainty in therefore in the results (Hillier and Lieberman, 2002). Some of these methodologies are the multiobjective analysis, multiobjective analysis with multiple decision-makers, decisions under certainty, decisions under risk, decisions under uncertainty, among others (Hillier and Lieberman, 2002).

The interest of this research is framed within the multiobjective analysis with a single decision maker, which is considered a very important research topic, not only because this type of problem is very common in real life, but because there are still many open questions in this area (Abbass et al., 2001).

The canonical paradigm of decision theory is characterized by following central elements (Bell, Raiffa and Tversky, 1988: 18; Marí-Klose,2000). We have, to begin with, an individual who has to make a decision anyone and from whom their preferences are taken for granted. The formal theory of the decision, as can be seen in the examples with which this book begins, does not come to consider the nature of the preferences of individuals - object of other disciplines - or why people prefer some things in others. From the formal perspective adopted by the theory, the only it is important that these preferences, whatever they may be, meet certain criteria of logical consistency, among which the the following:

Transitivity: for all x, y and z, if x is strictly preferred to and y is preferred strictly to z, x will be preferred to z.

Complete: For all x and all y, or x is preferred to y, or y is preferred to x, or the individual is indifferent between them

Asymmetry: If x is strictly preferred to y, y is not strictly preferred to x.

Symmetry of indifference: For all x and y, if x is indifferent to y, y is indifferent to x.

If all these requirements are violated at the same time it will be impossible to know what is which the person prefers; You can not sort out your preferences, and the decision theory will consider that person does not rationally choose, this is logically consistent Meet the requirement of transitivity assures us, for example, that we do not make a decision so that we leave disadvantaged, choosing at the beginning options that we prefer more - x front A and, e and against z- to end with a bad choice if we choose z versus x. The computes the person to compare all of his options with one another and to decide by one of them or manifest their indifference (which is a form of decision). To its asymmetry and symmetry are self-evident, and do not appear to impose an unreasonable logical exigency to the person who has to choose between several options: if we am indifferent between candidate A and B (or between car C and D), we can not say that we prefer candidate B to A (or car D to C); what if we prefer strictly candidate B to A, we will doubt my consistency if we affirm, at the same time, that we also prefer A to B. So, if these requirements are met the individual can be attributed a utility function, that is, an index or number to each of your preferences so that we can sort them from minor to higher, from least preferred to most preferred (Ríos, 1995).

In order to analyze the decision in these terms, it is necessary to identify previously a set of possible options from the perspective of who decision (its feasible set) and a set of consequences of each one of the options, consequences that can be anticipated and ordered according to the preferences of the individual. It is assumed that, given its feasible set, the individual will choose the option that has - or believes it to have - the best consequences, that is, the one that prefers more.

The decision can be parametric - if the context is considered given, it is to say, a parameter - or strategic, that is, if the decisions of the actors are interdependent, so that our decision depends on what the the rest. To this table should be added the amount of information with which account the individual to decide on an option or another of its feasible set. If the information on the results of the various options is complete - we know with certainty the consequences of our decisions – the individual will be faced with a situation of certainty; If, on the other hand, information is incomplete-we know what consequences our actions - the situation will either be at risk or very uncertain (later we will unravel in detail the difference between the two).

The parametric theory of decision (or decision theory, without more) addresses the formal nature of individual decisions and analyzes various criteria decision according to the information context in which the individual develops. This formal nature can be treated in a normative, prescriptive or descriptive way (Bell, Raiffa, Tversky, 1988; Selten, 1996: 289). In the first case, studies what decisions an idealized agent must take (which never suffers logical inconsistencies and that is able to optimize the search of information). The prescriptive theory of decision, however, deals with how they can choose real individuals, given their cognitive and informational limitations.

The descriptive theory of the decision studies how people actually decide. Game theory, in turn, analyzes individual decisions that are seen influenced not only by the contextual information available, but also by the decisions of others. It is, therefore, the formal study of strategic decisions, in which what a person decides depends on what information they have about what they do the others (Binmore, 1994). Game theory has played and plays a outstanding role in the recent development of disciplines such as sociology and politic science. Key sociological concepts such as social class, social norms or social structure are experiencing a remarkable development and foundation thanks to game theory (Aguiar, Criado and Herreros, 2003). In the same way, in the terrain of political science the strategic behavior of voters and parties, the problems posed by the control of politicians by citizens or international relations, among many other things, have taken on depth and perspective thanks to game theory (Morrov, 1994).

Finally, the theory of social choice studies and proposes criteria for add individual decision functions into a single social decision function or social choice function. Given the preferences of a set of people -for example, your preferences for different candidates in an election- it is a matter of knowing what their collective preference would be. This issue has given rise to a series of theorems of possibility and impossibility that show when it is possible, and when not, to add individual preferences through a rule of social decision (Arrow, 1974; Sen, 1976; Barbera, 1984). The contributions of the theory of social choice to the study of the different rules of decision in committees or electoral rules are now an inescapable aspect in the analysis of the functioning of democracy.

In this book we will deal exclusively with the theory of decision parametric in contexts of uncertainty, in its normative and descriptive versions. This is an aspect of decision theory that has had great influence in the social sciences, as well as in ethics and the theory of justice, since it is assumed to be in. The social and moral realm rarely have absolute certainty about the consequences of our actions. We will see first, then, some of the criteria that have been proposed in order to make decisions in situations of uncertainty. This will give, in the third section, to the theory of expected subjective utility, the most powerful mathematical tool, and more influential, to solve problems of uncertainty. Criticism of this theory, and in particular to his conception of rationality - so influential in different branches of the social sciences - will not have normative modelsto the descriptive ones, that have appeared in the last two decades as an alternative to the theory of expected subjective utility and whose influence has been left noted in sociology and political science.

RECENT METHODS OF MULTI-OBJECTIVE ANALYSIS

In the state of the art of multiobjective analysis can be found various methodologies that describe classic algorithms of optimization and that are very efficient and effective to the time to solve these kinds of problems. However, in the real world there are problems that classical methodologies can not solve easily, either because have extremely large search spaces that render the algorithm inefficient, because the approach makes use of nonlinear equations difficult to solve or because are simply not easy to pose mathematically and therefore can not be applied one of these methodologies.

The existence of such problems has led to explore metaheuristic techniques that offer the decision maker optimal solutions or at least close to them in the shortest time possible. Among the most recognized are Evolutionary Algorithms, Tabu Search, Simulated Annealing, Ant Colony, Particle Cloud, among others, which represent models from physics, biology or genetics.

In practice, one of the techniques most used to solve multiobjective problems have been the genetic algorithms, categorized in the family of evolutionary algorithms. Its high applicability is due to its efficient population search, since they work with several individuals simultaneously and find several solutions increasingly efficient and satisfactory in a single run.

Since 1967 with Rosenberg, genetic algorithms have been mentioned to solve multiobjective problems. Later, in 1983, researcher David

Schaffer raised a algorithm called VEGA (Schaffer, 1985) which was later criticized by David Golberg in 1989 that raises the need to make use of the concept of dominance to move the population towards the Pareto border. In the decade of the 90's many methodologies that sought to get closer to the real frontier in a way that efficient, such as MOGA (Fonseca and Fleming, 1993), NPGA (Horn et al., 1994), NSGA (Srinivas and Deb, 1994), SPEA (Zitzler and Thiele, 1999), among others. But in 2000 and 2001 NSGAII (Deb et al., 2002) and SPEA2 (Zitzler et al., 2001) greatly increased the effectiveness of this type of algorithm, and from there have been proposed different methodologies, although they have not been as well recognized by the scientific community as these of the latter.

However, with this type of methodologies, the problem of decision-making is not resolves completely, since in the end these offer the decision maker a fairly large number of alternatives that may be satisfactory to him, but do not offer him a single solution; So the decision maker must continue to solve a decision problem discreet considering your preferences.

In the state of the art (Coello, 2000), (Rachmawati and Srinivasan, 2006) the use of preferences not only at the end of the algorithm search, but before or during the search, in order to achieve satisfactory solutions for the decision-maker according to their preferences. However, it should be noted that the preferences of a decision maker are relative and changing, since they are subject to the environment they perceive in a moment therefore, it is risky to consider that the decision maker can decide preferences when you do not have prior knowledge of the problem.

This thesis intends to make a theoretical contribution both in the methodologies of algorithms multiobjective as in the use of preferences in an interactive way, since considers a very important topic to develop.

GENERAL AND SPECIFIC OBJECTIVES

The general objective is to make use of the strengths of interactive techniques to propose a methodology multiobjective based on genetic algorithms that allows the decision maker to intervene in effective way in the algorithm, so that, in the end, it gives a small number of solution alternatives close to the optimum and in the area of their interest.

Some of the specific objectives are given as follows:

- Propose a new methodology to solve multiobjective problems based on algorithms that improves the results obtained by existing techniques.
- Determine when the decision maker should intervene in the algorithm with preference, so that it does so in an efficient and practical way after be acquiring knowledge of the problem.
- Identify if it is necessary for the decision maker to modify parameters of the genetic algorithm while defining your preferences.
- Provide information to the decision maker, allowing him to acquire knowledge of the problem.
- Validate the methodologies proposed in a test problem by evaluating their efficiency using previously selected metrics and graphs.

DESISION-MAKING CRITERIA IN SITUATIONS OF UNCERTAINTY: THEORETICAL AND PRACTICAL RESULTS

The distinction between risk and uncertainty was established by F. Knight in 1921, who in his work „Risk, Uncertainty and Profit" referred to the first as that situation in which there is no certainty about the outcome of the decision, although at least the probability of different alternative outcomes is known. This would be the case, for example, of the choice between face or cross of a coin: we do not know beforehand the result (if the coin is not trucada, of course is) but we know the objective probability of the two alternatives. The situations uncertainty would be characterized, on the other hand, by the fact that only we do not know the final result, but we can not predict it either in terms of objective probabilities.

Thus, one of the central problems that the theory has faced of the decision has been to establish some criteria (or criteria) that allow us opt for one action or another in situations of uncertainty. Some of the more familiar criteria are set out below.

Criterion Maximin

Suppose we decide whether to make A or B, whose consequences are C, D and E, as reflected in the matrix of payments or results presented below. If we choose A we can obtain the most favorable result (100), but we also have two very bad results (2 and 1). If we choose B we risk not to win anything, although C and D do not offer both a good result. How we will decide what to do if we can not objectively attribute any probability to the consequences of our action.

	C	D	E
A	100	2	1
B	99	98	0

If we knew the probability of each of the events we would choose without further which will benefit us more safely. But as it is not, as we do not know the probabilities of events, what the maximin criterion advises us is to follow that action that assures us the maximum of the minimums; That is, the action which frees us from the worst possible outcome. This time it would be the action A that the maximin would recommend following, in the worst case, we would get as minimum 1, ie a better result than the worst of the results of action B. The maximin criterion compares the worst results of each of the options possible and choose the best of them (the maximum of the minimums). It is, therefore, of an essentially "conservative" criterion - it only seeks that we do not go very badly unemployed - that wastes much of the information that the matrix of payments, since it only takes into account the worst results of each.

Maximax Criterion

This criterion, unlike the previous one, only takes into it counts the best possible result of each action (the maximum of the maximum). In the previous example, action A would again be favored, since the best possible result is 100. Although it is not a strategy that seeks precisely avoid the risk at all costs - rather the opposite – suffers of the same defect as the maximin criterion: it does not take into most of the information that the situation of choice offers us.

Criterion of Hurwicz

To avoid the conservatism of the maximin and optimism of the maximax, Hurwicz proposed a criterion that weighted sum of the extreme results of both lines of action. Since no one is always optimistic or always pessimistic, the criterion of Hurwicz establishes a middle way between the maximin and the maximax. So, for example, if someone gives the worst results of A and B a value to = 3/4, and 1/4 (ie, 1-a) to the best results, the Hurwicz criterion will evaluate the different lines of action as follows:

$A=1\times3/4+100\times1/4=18,75$

$B=0\times3/4+99\times1/4=18$

Again, action A is chosen in our example. The choice of value thus, the degree of pessimism or optimism of the decision-maker equal to 1, the criterion of Hurwicz would be identical to the maximin; If it were equal to 0 would be identical to maximax. In any case, different decision makers will give different value a for purely subjective reasons. Thus, although this criterion does not preclude so much information as the previous two, it does not take it all into account, since excludes intermediate values.

Laplace's "Insufficient Reason" Criterion

The criterion of Laplace considers, instead, all values. Since we have no information on the probability of occurrence C, D or E, the most rational would be, according to this criterion, to assign to each value the same probability and choose the one that gives us the highest expected value (the highest expected utility). From this form would obtain the following results for A and B:

$A=1/3\times100+1/3\times2+1/3\times1=34,33$

$B=1/3\times99+1/3\times98+1/3\times0=65,33$

Unlike the other criteria, this is the only one that action A above A. This is undoubtedly due to the fact that take into account all values .

As we see, then, it is not only evident that there is no single criterion recommended to make decisions in uncertain situations, but, except for the Laplace, all appeal to the subjectivity of the one who decides: to his optimism or pessimism with respect to the situation of choice. Now, what happens if we approach decisions in probabilistic terms, such as the criterion of Laplace, but considering possible subjective attributions of that probability,as it is done in the Hurwicz criterion.

An answer to this question is provided by the Bayesian modification of the Laplace criterion. Instead of attributing the same objective probability to each of the outcomes, it would be possible for the decision-maker, for reasons believe that some events are more likely than others, attributing thus giving everyone a subjective probability. Bayesian modification would the subjective approximation of the maximin and the maximax and the probabilistic of Laplace. The mechanism would be similar to Hurwicz's criterion, but taking into account account all the information that gives us the structure of the decision problem, not just the extreme values. However, the attribution of probabilities by the decision maker to be subjective completely seems to confer on the process of arbitrary nature that detracts credibility and, therefore, decision itself. For although it is true that the degree of optimism or pessimism of the maximin and maximax criteria, for example, is also subjective – since being optimistic or pessimistic is something subjective-, the consequences of such subjectivity are limited to the exclusion of intermediate values, so that pessimists can bet on the action that reports them the best result of the worst, which is somewhat objective, and the optimists by which it provides them the best of the best (which is also something objective). That objective information from the structure of the decision is, in principle, absent from the mere attribution by the decision-maker of subjective probabilities to the different events, attribution that would be based on the observer's beliefs. Different observers will make different decisions, as they will attribute to probability events also different (French, 1986: 222). The undeniable halo of arbitrariness that this poses will be solved, as we will see in the section that follows from the theory of expected subjective utility (SEU model) ", which provides for the first time a theoretical model of great consistency and beauty to address the problem of decision in uncertain conditions.

Let us observe the following game: if we pay a certain amount of money to participate, we throw a coin to the air: if it comes face we paid 2 euros; If there is a cross, let's throw the currency, with the caveat that in this new launch we will have to pay double if it comes face. In this way, every time the reader loses, we will double the prize, and we will continue tossing the coin into the air until the reader wins, that is, until it comes face to face. The question is: how much would be willing to pay to play. The reader may reasonably believe, that a way of deciding how much they would pay to participate in the game can consist of adding up what can be gained by the probability of winning it, so that what is paid to play is the result (the expected value) of that amount. Well, since the probability of winning 2 euros in the first pitch is 1/2, that of winning 4 in the second is 1/4 (ie, the probability to lose in the first pitch and win in the second), and so on, the result of the sum will be as follows:

$(1/2\times2)+(1/4\times4)+(1/8\times8)+....=1+1+1+...=¥$

In this way it turns out that when opting for the rule of the expected value, wants to play will be willing to pay any price, however high. This, although mathematically impeccable, makes no sense, since nobody would pay amount exorbitant to take part in the game. What intuitively knows any player did not seem, however, mathematically demonstrable. It was the mathematician Daniel Bemoui (1700-1782) who provided a solution to this problem, which is known as the "St. Petersburg paradox" That as a person's wealth increases, the increase in income is less valuable, so it is more important to avoid a loss x of money that win jc. This preference can be represented by a function that project wealth on utility (which is the measure of desirability in numbers real), demonstrating in this way that if the function is logarithmic - that is, to a certain level of utility, where it begins to expected utility of the game will be finite. Thus, if the aim is to maximize the expected utility, will only be paid a moderate sum to enter the game.

The importance of this paradox and its solution lies not only in the fact that it provides a useful generalization to the rule of expected value, but demonstrates the compatibility between mathematical theory and subjective evaluations about uncertain future states.

We still have to wait, however, for the 20th century for these results to take body in the form of a theory capable of probabilistically measuring the individuals about the consequences of deciding on contexts of uncertainty. This work would be, fundamentally, done by L. Savage.

At the beginning of the thirties the philosopher and mathematician F. Ramsey had demonstrated that if individuals make their decisions under uncertainty as if they had to make a bet on what the possible outcomes would be its decision, it would be feasible to deduce the subjective probabilities of their bets". On the other hand, John von Neumann and Oskar Morgenstem proposed in the following decade a series of postulates of rationality to which the preferences of the person making a risky decision would have to be adjusted-in which the objective probability of events is known. The decision risky can be understood as a bet among different lotteries. From this in this way, a rational person will prefer, first, the lottery the biggest prize; Secondly, a rational person will be indifferent between a lottery whose only prizes are to win (G) or lose (P) and some lottery offer an intermediate reward between G and P. Thus, for example, if G is equal to 100 euros and P is equal to O, a person may interest you to pay 10 euros per participate in a bet that offers you the possibility of winning 50 euros with a probability of 5, but you will not pay 10 euros if your probability of winning is 0.01. There must therefore be a lottery that offers a prize between G and P and ante which bets are indifferent between playing that lottery or the lottery simple offering to win everything or lose everything.

From the works of Ramsey and von Neumann and Morgenstem, Savage demonstrated that our choices among different alternatives must be adjusted to certain conditions so that we can derive from them probabilities subjective and cardinal (ie, measurable) utilities. In other words, with the purpose of transforming the scarce information that we have in contexts uncertainty in an a priori distribution of subjective probabilities, the decision-maker has to consistently choose between different alternatives present in a series of simple hypothetical problems to solve, in which there remains implicitly reflected the structure of a possible real situation of choice. These conditions to which the choice must be made to be consistent, so that it is not entirely arbitrary, are the axioms of the SEU model, which we present below.

Axiom 1: Completeness.

Axiom 2: Uransitivity.

Axiom 3: Independence from Irrelevant Alternatives

For all actions x and y, the choice between both is determined only by preference over their respective consequences, and not in relation to the consequences of different alternatives. Let's see this with an example taken from Luce and Raiffa: In a restaurant the waiter tells us that only you have roast chicken or veal steak. We chose roasted chicken, but instantly returns the waiter apologizing because he has forgotten we offer trout, which also has left. After a few moments of doubt, we asked the waiter, since it also has trout, bring us veal fillet instead of chicken. A decision like this violates Axiom 3, since it is decided between two options having into account a third that is not relevant to the decision.

Axiom 4: Independence of counterfactual consequences

If two actions produce the consequences A and B, the preference by one action or another once one of them is performed will depend on the preference by its real consequences and not of others they could have.

Axiom 5: Independence from expected gain

If a decision maker has to choose between a bet or a different bet, the decision must based on the probability of winning and not on how much you can achieve.

Axiom 6: Minimal strict preference

There are at least a couple of consequences such that one is strictly preferred to the other.

Starting from these six axioms, Savage shows that the SEU model, based on the behavior of the decision maker, allows a) to assign subjective probabilities to the situations of choice, b) attribute cardinal profits to the consequences of the shares, c) calculate the expected profit of each associated bet with each alternative action and d) compare the different actions with each other numerically, choosing the one that maximizes the expected utility. "

One of the main features of Savage's theory is that it is not part of an objective conception of probabilities, but of its axioms is deduced the possibility of valuing the options in terms of subjective probabilities. This also allows us to attribute to the consequences cardinal utilities that reflect the preferences of decision makers. Another of the central and most discussed aspects of the SEU model is that it considers its axioms as canons of rationality. In other words, a decision maker facing a problem of uncertainty will be rational if and only if it maximizes their expected subjective utility, for which their preferences will have to be adapt to the axioms presented above. If this is not the case, the expected subjective utility, that is, the theorist who defends the SEU model, will consider that the decision maker is not rational. It is, as we see, a clearly normative, since the theory establishes what must be understood by rationality and, consequently, only the decision that conforms to the axioms of the model your is considered rational.

FROM THE PARADOX OF ST. PETERSBURG TO THE THEORY OF THE LIMITED RATIONALITY

Once the most essential of the SEU model is presented, let us now look at example how we would choose one option or another following this model. Suppose that we have decided by one of the following bets:

Bet A

a_1 : probability 1 to earn 1 million euros.

a_2 : probability 0.10 to earn 5 million, 0.89 to earn 1 million and 0.01 of not win anything.

Bet B

b_1, probability 0.11 to win 1 million and 0.89 to not win anything,

b_2 : 0.10 chance to win 5 million and 0.90 to win nothing.

Well, through a hypothetical example like this one can deduce the subjective probabilities of decision-makers and apply them to situations of choice of real life that have a structure similar to that of the example. What does it dictate? In this case, in this bet, the model SEU ?. According to Axioma 5, according to which the decision should be made considering the greater probability and not the quantity greater, the SEU model considers rational to choose a_1, and b_1.

However, the above example is not innocent, as it illustrates what is known as the paradox of Aliáis. Maurice Aliáis demonstrated in several articles that individuals systematically chose a, and b ^, which contradicts what is required by the SEU model of decision under uncertainty .

Since Savage proposed his theory in the 1950s, there is no doubt that which has experienced enormous success and development, but at the same time have multiplied experiments with the SEU model that demonstrate that people violates his axioms with great frequency. Moreover, as evidenced by Tversky and Kahneman in numerous articles, the intuition of people in relation to the calculation of probabilities is very poor. This leads to multiple biases in making decisions under uncertainty. It is not uncommon, for example, that whoever has to make a decision falls into the so-called fallacy of conjunction: A and B are more likely to occur at the same time to be given only A, for example. Likewise, it is also frequent to be a victim of and the player's fallacy, who has not said to himself ever "we will follow playing, we'm on the run, "or" since we've had two good runs we'm leaving because has to come a bad one? Examples similar to these are many ".

Does it imply the above, perhaps, that people are irrational? Maybe yes yes We consider that the SEU model represents how we should make decisions rational In other words, if it is stated that the theory of expected utility subjective is a normative theory - which tells us how we should decide - and not descriptive - which would tell us how we decided in fact - it could be argued that any decision that departs from the SEU axioms is irrational. However, today in day nobody considers that the violation of the SEU model implies the irrationality from the people. Rather, it has sought to relax the assumptions of the model to better fit the way people make their decisions. In both decades, many developments have emerged from the SEU model that relax or reject some of the axioms we saw above, especially the three first .

However, some people consider that it is not enough to touch up the SEU model, but that, insofar as it does not describe the actual behavior of people,it is wrong and must be replaced by another. Simon's proposal implies a triple transformation of the SEU model, since this author advocates, first place, by a procedural conception of rationality instead of conception substantive of the SEU; Secondly, it replaces the concept of maximization by the one of satisfaction (that is, Simón assures that the decision maker does not worry both by choosing the optimal and by choosing an action whose result please). Finally, these transformations lead to a descriptive theory of the decision regarding the normative character of the SEU. In his own words Simon,

The task is to replace the global rationality of the economic man by a type of rational behavior that is compatible with access to the information and with the computational capacities that they actually possess organisms, including man, in those contexts where such organisms (Simon, 1957)

SEU theory prescribes unique solutions for decision problems under uncertainty, but "does not offer a procedure to find that solution with an acceptable amount of computational effort "(Simon, 1986). Rather, Savage's axioms assume that individuals have a computational ability and unlimited information. But this is not so. Well, first of all, our decisions do not usually affect, according to Simon, to "great areas" of our life, but, rather, very specific problems, independent of each other. Second, almost nobody makes a decision, even if it is important, elaborating hypothetical situations to which they attribute a distribution of subjective probabilities. In addition, the de facto to have to decide for one thing and not for another (buy a car and not a house), makes us focus on some aspects of the decision and not in other (focalpoints), and in some of our values and not in others. Finally, most of our effort in making a decision we devote it to collecting information. A theory of the decision we take in consideration all these elements will be a theory, according to Simon, of rationality bounded rationality: the work of this theory is to describe the procedures by which individuals make their decisions (Simon, 1983). Hence, this author proposes a procedural concept of rationality against the substantive (axiomatic) criterion of the SEU model, and from there which also advocates a descriptive theory of the decision based on the limitations of rationality (Simen, 1955, 1983, 1987, 1992, Rubinstein, 1998; Gigerenzer and Selten, 2000; Todd and Gigerenzer, 2003).

Assuming the limitations of human rationality in contexts of uncertainty implies that the axioms of the SEU model can not be met. In special, for Simon it is not possible to make complete ordinations of our preferences. This would imply an enormous computational capacity on the part of the individuals: the most we can aspire to is to establish partial ordinations, because we will not always be able to compare each and every pair of the options that are presented to us, so that we can decide for a or manifest our indifference (there will be times when the comparison is impossible). The consequence of this is that the maximization of expected utility it's not possible. Without completeness there can be no maximization. What proposes Simón is to replace the scalar pay-offs of the SEU model with vector payment function V , Where V has different components Vj, V_2, ... V_s. " In this way, rather than choosing that value that maximizes our utility expected, what we do is to determine the vector that contains a set of payments that satisfy us, as shown in Figure we (Simon, 1955: 108-109).

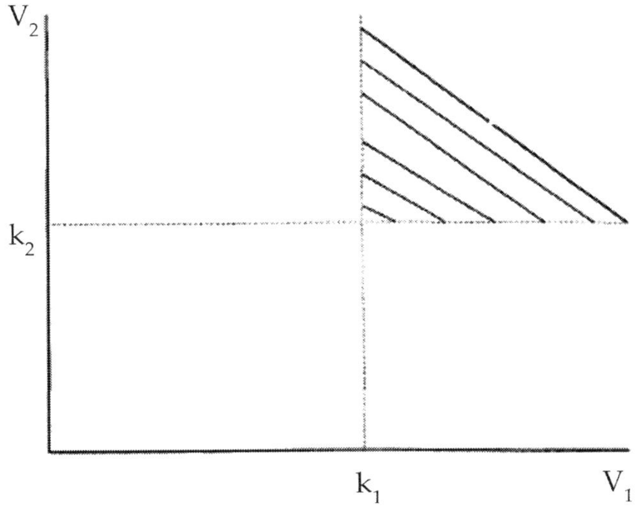

Region of Satisfactory Payments.

So, for example, if we want to buy a house we will not necessarily buy that which maximizes my usefulness (which may be impossible), but that we will compare limited set of houses that for their price and characteristics satisfy me (Although none of them maximize my usefulness in all possible aspects, price, size, environment, distribution, etc.). What is lost, then, in mathematical precision is undoubtedly won by the realism of the model.

However, we can not assume that in any context of uncertainty the decision maker always follows a rule of satisfaction rather than maximizing its expected utility. Most likely, different situations of choice under uncertainty, require different rules of decision. As have shown by Payne, Bettman and Johnson (1993) -to which we can frame in the school of limited rationality - people not only decide for a option or another, but decide how to decide. Thus, against theories of decision under uncertainty that try to find the rule of rational decision (maximin, SEU, satisfaction ...), what we would have, according to these authors, is a decision maker who uses different rules, thus adapting to the greater or lesser difficulties of the choice. The basic assumptions of the adaptive decision-maker theory are therefore the (Payne, Bettman and Johnson, 1993: 14):

- Decisions are sequences of mental operations that can be represented as operators of the form
- IF (CONDITION 1, ..., CONDITION N) THEN (ACTION we, ... ACTION N).
- The cognitive effort required to make a decision is in function of the number and type of operators, which will be in turn in relation with the context in which it is chosen.
- Different decision rules are characterized by different levels of precision, being the levels of precision of the contingent rules in relation to the context. Precisely defined is the choice of a "good decision ", that is, that provides a good result using the relevant information.
- Decision-makers are assumed to have more than one decision rule (sequence of operations) according to their training and previous experience.
- Individuals decide how to decide considering both cognitive effort as well as the precision of the different decision rules.
- In the selection of the decision rule can influence factors such as need to justify to others the decision or to minimize the conflict which may be a decision problem.
- The decision of how to decide is conscious sometimes and others learned.
- The selection of the decision rule is adaptive and intelligent, if not optimal.

Based on these assumptions, the theory of decision must investigate the existence of different decision rules, as well as their precision and cognitive effort which they suppose in terms of information. In this way, we will have rules that require a great deal of cognitive effort and others that require much less (And, perhaps, provide a less accurate result, which will be determined Largely because of the context in which they are applied).

The additive rule which is weighted, for example, is very demanding as far as evaluates all the alternatives according to each and every one of its attributes relevantes. So, for example, if we want to buy a car we will have to compare, given my budget, the whole range of possible cars, each and every one of its characteristics, order according to their importance these characteristics, check the different prices of all dealers, etc. This would be the case, for example, of the SEU model of decision, which is criticized precisely because unrealistic in terms of the amount of information it requires to be able to decide. The equivalent weight rule also examines all alternatives and all the attributes of the same, but simplifies the decision by not having in account the information about the relative importance of each attribute. By last, the rule of elimination by aspects identifies, first, the attribute most valued among all the alternatives (for example, that the engine of the car we want to buy is turbodiesel). Then delete those alternatives that possess this attribute do not reach the required value or quality. The process continues with the second most valued attribute (consumption gasoline, for example), then the third, and so on until only there is an alternative.

The response of the defenders of the SEU model to the criticisms of descriptive theories of the decision implies the acceptance that the rationality of people is limited, but they are considered to make their decisions as if they had a unlimited rationality, that is, an unlimited computing capacity of the decisions. The development of ecological, adaptive or evolutionary models of decision-seeking to be more realistic, as they try to describe how they take people make their decisions and place great emphasis on learning and the trial and error - questions such defense of the SEU model. In fact, it seems already sufficiently demonstrated by experimental psychology and economics that people assume the limitations of rationality and try to adapt to the circumstances of the election (difficulty of the election, time to importance of the problem, etc.), using simple rules of decision and 'Tricks' to save yourself the search for new information or to make decisions when it is lacking (Gigerenzer, Todd et al., 1999, Gigerenzer and Selten, 2002). Thus, the perspective of ecological rationality considers that if rationality is limited not only to the limitations of the human mind or to the difficult to obtain information in complex contexts, but to the sum of both factors.

Given the cognitive limitations of people, the ecological rationality tries to find out what rules of nature decision heuristic uses the human mind to adapt to its environment in the best way possible. It can not therefore be assumed that people make decisions as if their capacity unlimited, but quite the contrary. People, through adaptive rules that make the decision quick and inexpensive as far as the search for information, try to make the most of the structure of the context or environment in which they move. Thus, under the appellation of fast and frugal heuristic rules of decision heuristics are the multiple criteria or decision rules used really the people in the most diverse contexts to make little decisions costly and effective (Todd and Gigerenzer, 2003). If, for example, we do not know well the problem on which we have to decide, if it is impossible to achieve sufficient information or if the alternatives are presented sequentially we use different heuristic rules capable of taking advantage of our cognitive implications as contextual difficulties. Heuristic rules would be, therefore, a "collection of specialized cognitive mechanisms that evolution and learning have built in our minds "(Todd and Gigerenzer, 2003).

Two examples, to end this section, may help to better understand the nature of heuristic rules than the theory of rationality considers that people use when making real decisions. Among the most studied heuristic decision criteria, it should be noted that are known as Take The Last (TTL) and Take The Best (TTB). If we count on little TTL information advises us to use the latest, the latest, in function of the context in which we find ourselves. Thus, for example, if someone had to choose which three or four decision theory articles to read from all how many have published between 1950 and 2004, the most reasonable would be to start with the latter. This is precisely what is usually done. The SEU model would, on the other hand, to compare all articles with each other, the authors, their influence, etc., before to have read them, to choose which ones to read. For its part, experimental studies have shown that people use TTB when they lack information to discriminate between two or more options. TTB recommends that we choose the attribute that, for whatever reasons, convince us more and forget the rest (take the best andforget the rest) (Robles, 2004).

If we want to read three articles of theory of the decision and we do not have a determining criterion that is not informative very expensive we can choose the most recent (TTL) of American authors (TTB), for example, suppose that the five most recent articles, all from the same year, they are American authors. To choose three of those five we select then the top three universities (TTB) and set aside another type of criteria. The theory of ecological rationality does not ignore that with these rules Heuristics can make mistakes when deciding, that is, they occur inferential failures leading to poor outcomes (Broder, 2000). But the question more important is not that heuristics can cause failures, but to demonstrate in a certain way that the real people adopt this type of heuristic in making their decisions. This requires much more empirical work.

Although there is no doubt that the problem is still in the process of development and the one that still has a greater number of applications (and followers) in fields as diverse as economics, sociology or ethics, we have attended in the last fifteen years to a great development of decision models alternatives. In some cases this alternative is reduced to cancellation or review some of its axioms.

In others, as in the theory of limited rationality of Simon and his followers, a whole model is proposed alternative that has more and more weight in sciences such as sociology or politics. In any case, it seems that the general trend of decision theory consists in giving up much of the mathematical precision and parsimony methodological model of the SEU model in favor of greater realism of assumptions Realism supported by decades of experimental work with the model of expected subjective utility in which it has been shown that people violate their axioms without being irrational. The question that open is whether the social sciences that rely on the SEU model – all the currents that are identified with the theory of rational choice – must based on the new descriptive models or will continue to opt for a model normative of individual behavior. There is no doubt that the this question will determine the future of some of the theoretical currents most important in social sciences.

CHAPTER 2

OVERVIEW OF DECISION-MAKING METHODS IN MULTI-OBJECTIVE PROBLEMS

Ivan Stanimirović

INTRODUCTION

In the decision-making processes in the real world under the concept of optimization, it was necessary to take into account several objectives simultaneously. In this type of problem is not a single solution, but a set of solutions that are optimal or close to being.

Given that a single solution must finally be chosen, the decision-maker must make use of its preferences with respect to the criteria being solution that best fits your interests. Depending on how you are interested to involve their preferences, whether before, during or after the optimization process, the decision maker will be able to solve his multiobjective problem with a specific technique.

Within the multiobjective optimization there are problems where classic techniques difficult to find a good solution, these kinds of problems have been dealt with meta-heuristic techniques finding satisfactory solutions for the decision maker.

In the next section we will describe the mathematical definition of a problemn of multiobjective optimization, then in section 2.3 we will mention some of the techniques used to solve this type of problem taking into account the moment in that the decision maker involves their preferences. Section 2.4 lists two types of difficult to solve with classical techniques, and have been treated with techniques meta-heuristics, which are mentioned in section 2.5. Finally, a specific meta-heuristic technique called Evolutionary algorithm is defined, to which belong the genetic algorithms which are the object of this investigation.

MULTI-OBJECTIVE OPTIMIZATION PROBLEMS

For many years the need to include in the intake process has been considered. Of decisions under the optimization scheme, several objectives simultaneously, which may not have the same units with each other. This need has led to the need to make use of tools that allow simultaneous inclusion of multiple objectives that can be represented mathematically or through qualitative expressions (Smith et al., 2000).

Problems where they are considered more than one objective are called Problems Multiobjective, and its purpose is to optimize all the objectives simultaneously, which, most of the time, are in conflicts. When you talk about optimality, refers to finding the best possible solution or a good approximation of this, given a set of limitations or restrictions. A multiobjective problem is represented mathematically as follows (Abbass et al., 2001):

Optimize $F\left(\vec{x}\right)$ subject to:

$$\Omega = \{\vec{x} \in R^{*} \left| G\left(\vec{x}\right) \leq 0\right\}$$

where $\vec{x}$ is a vector of decision variables $(x_1, ..., x_n)$ and $F\left(\vec{x}\right)$ is a vector of functions objective. The functions $(f_1(x^*), \ldots f_m(x^n))$ belong to R^n and

Ω is a non-empty set belonging to R^n. The vector $G(\vec{x})$ represents the restrictions that must meet the solution, these may be greater or less than zero (Abbass et al. 2001).

The purpose of multiobjective problems is to find a vector

$\vec{x}^*$

belonging to the region Ω and that optimizes $F(\vec{x})$, where each of the functions can be maximized or minimized, depending on the problem. A multiobjective problem does not have a single efficient solution, rather, it has a set of efficient solutions that can not be considered different from each other. This set of solutions is called Pareto Border.

A solution is efficient when it is not mastered, that is, when it is at least as good like the others in all its objectives and is better in at least one of them (Abbass et al. 2001).

The set of solutions belonging to the Pareto S boundary, in a problem where all its objectives need to be maximized, can be represented by the following manner (Smith et al., 2000):

$S = \{\vec{x} : \vec{x} \in \vec{X}$, there is no other $\vec{x}^* \in \vec{X}$ such that:

$f_i\left(\vec{x}^*\right) > f_i\left(\vec{x}\right)$ for all $i \in \{1, 2, ..., m\}$ and $f_k\left(\vec{x}^*\right) \geq f_k\left(\vec{x}\right)$ { for k≠i} where m represents the number of objectives considered, $f_i(x^*)$ is the value of the objective i for solution $\vec{x}$,..., and $\vec{X}$ is the space of possible solutions. Therefore, if one wishes to change from one uncontrolled solution to another, within the space of the objectives, in order to improve the value of one of them, it is necessary that one or more objectives should become worse (Smith et al., 2000).

In the case of a problem where we want to maximize two objectives, the which can be drawn on a Cartesian plane, the Pareto border can be represented as shown in Figure 1 (Smith et al., 2000).

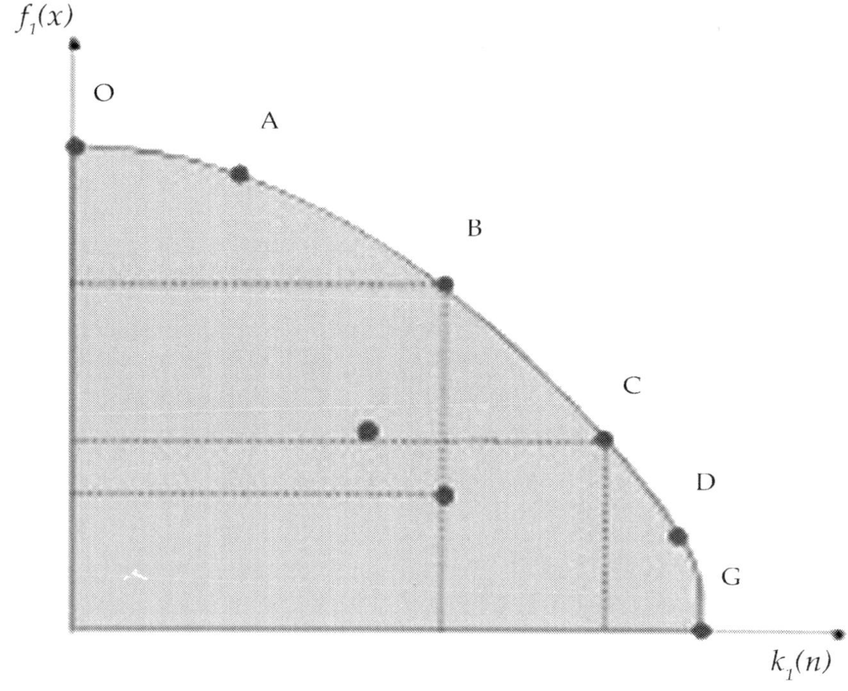

Figure 1. Representation of the Pareto Border taken from (Smith et al.2000).

The shaded area is the space of possible solutions and non-dominated solutions are represented in the OABCDG curve, called the transformation curve. In this case, point B has a higher value in objective 2 than point C, but lower value for objective 1, therefore it is not possible to say that a point is better than the other. If you observe, for example, the points F and E that are not inside the curve, it can be noted that they are always dominated by one or more points, for example point has the same value in objective 1 with respect to point B, but has a lower value of objective 2, whereby E is said to be dominated by B.

Because in a real problem usually only need to implement a solution, decision-making process must necessarily involve a decision-maker or a group of them choosing one among a set of possible efficient alternatives. For this investigation will only take into account the case where a single person as a decision maker. This person always has a preference structure where represents how important an objective for him is in relation to the rest. This structure always influences when making a decision, since the decision maker finally selects the alternative, among all possible, that produces the consequences with greater preference for him.

The preference structure of the decision maker implies a degree of difficulty, since one decision simultaneously affects several attributes that possibly compete with each other, which means that if the decision taken can fully satisfy the preference of the decider in a given attribute, it is very likely that it will not be fulfilling for another attribute (Smith et al., 2000). This makes for the decision maker it is important that the technique used to solve its decision problem contains in one form or another the representation of your preferences to feel at ease with your result.

To mention some of the techniques used to solve multiobjective problems, use will be made of the classification proposed by Cohon and Marks in 1975 (Cohon and Marks,1975), which classify them in A priori, Interactive or A posteriori techniques. In the first group preferences are involved at the beginning of the process; In the second they are involved as the algorithm progresses; And in the latter they become involved when you have a group of solutions.

TECHNIQUES FOR SOLVING MULTI-OBJECTIVE PROBLEMS

A-Priori Techniques

Techniques classified under this approach require the decision maker to represent their preferences before starting the search process for the solution. These may be represented by means of a desired solution, a proposed goal or, for example, a function that represents its structure of preference.

The fact of having to define the preferences before starting the process, makes for the decision maker is a bit difficult since it does not know the behavior of the problem. By example, in the case of using a technique where the decision maker must define his goals to fulfill, the latter could define them as not optimistic and achieve less satisfactory results than if it set a more ambitious goal.

Some of the techniques that are classified within this group are:

Schedule by Commitment

This method was initially proposed by Zeleny (1973), who proposed to find a solution or set of them that are closer to an ideal point proposed by the decision maker. This point represents what the decision-maker wants to achieve in each objective and therefore general, due to the conflict between the objectives, is an unreachable point. The set of solutions closest to this ideal point, called set commitment, distance indicator, which in the case of two represents by means of the following equation:

$$L_1 \to MinL_1 = \left[\sum_{i=1}^{m} w_i \frac{f_i^* - f_i(x)}{f_i^* - f_{*i}} \right]$$

And when you have more than two objectives and you want to have a solution that represents greater equity between them, in the previous formula appears a new infinite parameter, the distance indicator being the following:

$$L_{\max} \to MinL_{\max} = d$$

Subject to:

$$w_i \frac{f_i^* - f_i(x)}{f_i^* - f_{*i}} \le d \quad 1 \le i \le m$$

where w_i represents the weighting factors of the objectives,

- The ideal point and
- The anti-ideal point.

There are other distance metrics, the most common being (Smith et al., 2000):

$$L_k = \left\{ \sum_{i=1}^{m} w_i^k [f_{i,Max} - f_i(x)]^k \right\}^{1/k}$$

Where represents the relative importance of the objective? And k the degree of the meter

(1 <k <infinity) reflecting the importance of maximum deviation. These

two values must be defined by the decision maker and therefore represent their preferences. If the objectives are not expressed in the same unit, then these must be rescaled and for this the metric k L can be transformed into (Smith et al., 2000):

$$Min\left\{ L_k(x) = \sum_{i=1}^{m} w_i^k \left\{ \frac{[f_{i,Max} - f_i(x)]}{[f_{i,Max} - f_{i,Min}]} \right\}^k \right\}$$

This methodology can be consulted in Zeleny (1973) or in Smith et al (2000).

Goal Programming

It has been one of the most widely used methods in multiobjective analysis. It was developed has been one of the most widely used methods in multiobjective analysis. It was developed by Charnes, Cooper and Ferguson (1955) and aims to minimize the level of methodology for multi-objective optimization based on GA and use of preferences overview of multiobjective problems deviation between the achieved goal of the objective and a goal originally proposed by thedecision maker.

The formulation of goal programming is expressed as follows (Smith et al., 2000):

$$Min\left\{ \sum_{i=1}^{m} \left| f_{i,Max} - f_i(x) \right| \right\} \quad x \in X$$

In order to convert the previous equation into a linear function two new variables for each objective d_i^+ and d_i^- which correspond to the positive deviation (Excess) and to the negative deviation (deficit) of $f_i(x)$ with respect $f_{i,Max}$ to goal f, respectively. With these new variables, the problem becomes:

$$Min\left\{ \sum_{i=1}^{m} (d_i^+ + d_i^-) \right\}$$

Subject to:

$$f_i(x) + d_i^+ - d_i^- = f_{i,Max} \qquad \forall i = 1, 2, \ldots, m$$

$$x \in X$$

For further study of this methodology, Charnes et al. (1955) or Smith et al. (2000).

Utility Functions Or Multi-Attribute Value

A utility or individual value function is a function that converts the value of a objective or attribute in another that represents the preference of the individual in relation to that value.

The value function differs from those of utility, because the former are defined in a deterministic context and the second in a probabilistic one. In the case of the functions of value, the decision-maker will define it taking into account the importance it gives to the function objective that it is evaluating, instead that the utility function constructs it determining the preferences of the decision maker before several scenarios.

Find the utility or value functions for each objective, you can find the function utility or multi-attribute value. This is achieved through an aggregation of individual functions of each objective. If one considers independence between individual functions, this aggregation can be defined from the following rules:

Additive Rule:

$$U(G) = \sum_{i=1}^{m} k_i u_i(G_i)$$

Multiplicative Rule:

$$1 + kU(G) = \prod_{i=1}^{m} [1 + kk_i(G_i)]$$

Where U(G) is the utility function or multi-attribute value that takes values between zero and one, the k_i where i, = 1,2 ..m are positive constants and less than 1 given by the decision maker and which represent the preferences of this, k is a constant that satisfies the following equation:

$$1+k=\prod_{i=1}^{m}[1+kk_i]$$

For further study of this methodology (Smith et al.2000).

Interactive Techniques

The techniques that are classified in this group are those where the preferences of the decision maker are expressed interactively during the run of the algorithm, that is, they allow the decision maker to restructure their preferences if they wish, after knowing the solutions partially unmasked.

Depending on the methodology, the decision-maker provides one type of information in each iteration, for example, can provide the tradeoffs or exchange rates between the objectives after knowing the objective values in each iteration; You can select one alternative from knowing an efficient small set of them and the value of their objectives or can provide new benchmarks according to their preferences after knowing the values reached in each objective function (Caballero et al., 2002).

Each of these methods has the advantage that the decision maker has the possibility to calibrate their preferences during the process, since they often do not have clearly defined. However, depending on how you are asked information to the decision maker in each iteration, the method may require a great effort mental to the decision maker. Some of the techniques that are classified within this group are the following

Stem Method

This method was published by the authors R. Benayoun, J. De Montgolfier, J. Tergny and O. Laritchev (1971) and consists in restricting the set of opportunities in each iteration, depending on the preferences expressed by the decision maker. Preferences represent a satisfactory goal, in order to improve it still is not.

A solution is obtained after minimizing the distance of Tchebychev from the vector ideal on the restricted set of opportunities.

The algorithm consists of the following (Caballero et al., 2002):

- The ideals f_i* where i= 1, ..m, and anti-ideals f_i=............i =1, ...,m
- The values of the parameters π_i (normalizing weights) are obtained

$$\pi_i = \begin{cases} \dfrac{f_i^* - f_{i*}}{f_i^*}\left(\sum_{j=1}^{n} c_{ij}^{\ 2}\right)^{-1/2} & f_i^* < 0 \\ \dfrac{f_i^* - f_{i*}}{f_{i*}}\left(\sum_{j=1}^{n} c_{ij}^{\ 2}\right)^{-1/2} & f_i^* \geq 0 \end{cases} \qquad i = 1, 2, \ldots, m$$

where c_{ij} are the coefficients of the objective function f_j and it is denoted by J^h set indexes of the functions to be relaxed at iteration h and initialize $X^0 = X$, $J^0 = Æ^F$, J = 0

- The weights of the distance of the metric of Tchebychev are calculated:

$$\mu_i^k = \begin{cases} 0 & si \ \ i \in J^k \\ \dfrac{\pi_i}{\sum_{j^L J^h} \pi_j} & si \ \ i \in \{1, \ldots, m\} - J^h \end{cases}$$

- The solution is obtained that minimizes the distance from the ideal to the target space restricted $Z^h = f(X^h)$. To do this, the problem is solved:

$$\min_{x,\alpha}$$

$$\mu_i \,(f_i(X) - f_i^*) \leq \alpha \qquad i = 1, \ldots, m$$

$$x \in X^h \qquad \alpha \geq 0$$

This solution is denoted by (x^h, f^h) and its corresponding criterion vector $f^h = f(x^h)$.

- The decision maker expresses preferences about the solution:

If it is satisfied, it ends with (x^h, f^h) as the final solution. Otherwise, keep going.

- If continuous is taken $h = h + 1$. The decision-maker specifies:
- Functions to improve (f_i ie{1,...,m}-J^h)
- Functions to relax (f_i ieJ^h), with the maximum amounts to be relaxed (Df_i^h ieJ^h)

In this way we have the feasible region of the following iteration:

$$X^h = x \in X / \begin{cases} f_i(x) \leq f_i\ (x^{h-1}) + \Delta f_i^h & i \in J^k \\ f_i(x) \leq f_i\ (x^{h-1}) & i \in \{1,\ldots,m\} - J^h \end{cases}$$

- The distance weights of the Tchebychev metric are recalculated (back to step 3), note that $x^{h-1}eX^h$.

If further research on this methodology is desired, Benayoun et al. to the. (1971) or Caballero et al. (2002).

Visual Interactive Approach Method – VIA

This method was published by Korhonen and Laakso (1986), uses a normalized function achievement, which should be minimized according to the benchmark provided by the decision maker at each iteration. In addition to defining the reference point, the decision-maker should to choose between several solutions (Caballero et al., 2002). The method is described below (Caballero et al., 2002):

- Let h = 0. The decider sets the weights (w_1, ..., w_m) for the normalized function of achievement. A feasible initial solution is obtained by solving the problem:

$$\min_{x,\alpha} \alpha$$

$$w_i\ (f_i(x) - f_i^*) \leq \alpha \qquad i \leq i \leq m$$

$$x \in X^h$$

where f_i^* is the ideal or reference point.

- The decision maker is asked if he is satisfied with the solution obtained, if the answer is positive, then the process ends and the final solution is (x^h, f^k). And if the answer is negative, then h = h +1 and is continued.

- The decision maker is asked to specify new goals or benchmark $\overline{q}^h$, taking into account the values obtained in the objective functions.
- For different values of λ in the interval [0,1], including l=0, the problem is solved:

$$\min_{x,\alpha} \alpha$$

$$w_i\,(f_i(x) - (\overline{q}_i^h + \lambda(f_i^{h-1} - \overline{q}_i^h))) \leq \alpha \qquad i \leq i \leq m$$

$$x \in X^h$$

Obtaining the solution criterion vectors denoted by $\{f^{h,0}, f^{h,11}, \ldots f^{h1,11}\}$

- From all these vectors criterion $\{f^{h,0}, f^{h,11}, \ldots f^{h1,11}\}$, the decision maker chooses the most preferred, denoted by h f and its corresponding vector in the decision space by x^k. And go back to step 2.

To see a practical example of this methodology can be consulted (Caballero et al.2002).

Equalizer Method

It is a method proposed by Jaramillo (1999) and implemented in a support system of decision. This method consists of refining an initial solution, obtained by others methods of multiobjective analysis without need to express the preferences of the decision maker. The decision maker does this refining, knowing in advance the consequences of his possible actions, and after offering you information that allows you to understand relations in the decision space of the problem, such as limitations, feasibility, efficient exchanges, etc. (Smith et al., 2000).

The following is a description of how this methodology works (Smith et al.2000):

- Calculate an initial solution by the method and weights that the decision maker chooses (programming of commitment with metric one, two and infinite and programming by goals). The solution is shown to the decision maker by means of a graph in which it presents movable vertical bars, one for each objective, which describe for each objective the best possible value (independent of the other objectives), the worst value possible and the value of the efficient alternative obtained initially.

- The decision maker has the option to keep this solution or register it for analysis later.
- The decision-maker decides what goal he wants to improve at the expense of someone else and where he is willing to worsen the latter.
- The program immediately, as much as possible, improves the goal you chose and shows the efficient and feasible solution that meets these conditions. The solution is obtained by solving the following mathematical programming problem:

$$\max f_k(x)$$

$$f_j(x) \geq \in_j$$

$$f_i(x) \geq \in_i^{'} \; \forall i \neq k, i \neq j, i = 1, 2, \ldots, m$$

$$x \in X$$

where k is the target chosen for improvement, j is the target chosen to sacrifice up to level and $\in_j \; y \in_i^{'}$ represents the values of $f_i(x)$ i of the previous solution of the equalizer, corresponding to the levels marked on the movable bars of each of the objectives i before performing the optimization.

- The decision maker has the option to register this solution and continue his search and go registering the solutions that you find interesting, or if you want you can stay with the group found here and continue with the next step.
- The user evaluates whether the saved group of alternatives is still interesting to him or contrary there are some alternatives that have already lost their interest and therefore eliminate them

from the register. If removing these alternatives still leaves a group of them, they are evaluate with some discreet method proposed by the tool. To go deeper into this methodology, one can refer to (Jaramillo 1999).

A-Posteriori Techniques

This group of techniques does not require the decision maker to express their preferences before, nor during the algorithm, they are limited to finding the Pareto frontier or solutions not dominated by modifying some of its parameters. Each methodology is proposed in order to find the whole Pareto front by modifying some of its parameters without having to express their preferences, and only up to the end, the decision maker could choose which of the solutions found would be most satisfied according to their priorities, so this would imply that before taking a each algorithm must be run as many times as possible as border is desired.

Within this group could be mentioned some techniques as the following.

Weighting Methods

It is a method proposed by Zadeh (1963) and is to obtain for a problem a set of non-dominated alternatives, without establishing an order of preferences. Zadeh showed that optimizing the weighted sum of the functions objectives is generated a non-dominated point, that is why it proposes to convert a multiobjective problem in a monobjective, as follows:

$$Max\left\{w_1\frac{f_1(x)}{f_1^{'}}+w_2\frac{f_2(x)}{f_2^{'}}....+w_l\frac{f_l(x)}{f_l^{'}}....+w_m\frac{f_m(x)}{f_m^{'}}\right\}$$

$$ax \in X$$

The constraints are the same as the original problem. If f_i'is the normalizing constant the objective i = in order to be able to sum them together and w_i = is the weight factor of objective i. For each of the possible combinations of w_i must be satisfied that:

$$\sum_{i=1}^{m} w_i = 1$$

$$0 \leq w_i \leq 1$$

With a combination of i w we obtain a non-dominated solution, to obtain another this combination must be varied. More on this methodology can be found in Zadeh (1963) or(Smith et al.2000).

Method of Restrictions

This method was proposed by Marglin (1967), and proposes to convert a problem multiobjective in a mono-objective by choosing one of the objective functions to be optimized and the others are handled as constraints. So:

Max $f_k(x)$

$f_i(x)^3 \varepsilon_i$ i k

xeX

where $f_k(x)$ is chosen arbitrarily as the objective function to be optimized.

A correct variation ε_i of the in order to obtain an adequate number solutions, in addition to knowing the minimum and maximum limits for each objective and thus know among which values should be varied ε_i. Under this same concept, several researchers have proposed different algorithms that initially find the allowed range of the ε_i and then generate the set of solutions of the problem under study.

For more on this methodology, see Marglin (1967) or (Smith et al. 2000).

MULTI-OBJECTIVE PROBLEMS DIFFICULT TO RESOLVE

Within the real problems that consider multiple objectives, there is a set that difficulties in resolving, irrespective of the use to resolve them. One of these are nonlinear and combinatorial problems.

The former are those where their objective functions or constraints can not be represented by linear form, but on the contrary are represented by non-linear functions and that therefore are very difficult to solve with a classic technique. The second are those which have many decision variables and these take values of 0 or 1.

In this investigation, we will take a problem of the second type.

Nonlinear Problems

There are real problems both monobjective and multiobjective, where their objectives and constraints it is not possible to approximate them to linear functions, so it is necessary to use of nonlinear equations and solve them using non-linear programming models linear.

This type of programming has complexities such as having to consider that the solution will not necessarily find it at the ends of the feasible region such as the case of linear programming, and therefore must consider all solutions within this region. In addition, it is possible to find a solution as optimal, but in reality is local and not global, so non-linear programming algorithms must be careful not to get caught up in a solution of these (Hillier and Hillier, 2002).

Combinatorial Problems

There are cases, in monobjective and multiobjective programming, that although the space of the decision variables is discrete, there is no small group of alternatives possible, but this is quite numerous or even infinite, when this occurs has an optimization problem defined as combinatorial. In it the options usually be all possible combinations of all possible values of the variables of constraints (Smith et al., 2000). Some typical examples combinatorial are the scheduling of tasks (production process of a factory), selection of routes to follow (problem of the traveler), problems of sequential assembly (Order to place parts in the manufacture of an automobile), selection of projects, among others.

This type of problems, usually are NP-Hard, that is, the time used by the algorithm to solve it increases considerably with just increasing in one the number of variables (Smith et al., 2000), and are difficult to address with classical techniques of mathematical programming.

For this type of situations and for the cases in which the problem to be optimized is mathematically complex with non-linear functions, another kind of methods known as metaheuristics.

META-HEURISTIC TECHNIQUES

These methods, also called intelligent systems by authors coming from the area of Artificial Intelligence, consist of systematizing ideas to develop algorithms efficient solutions that find "good solutions" to monobjective optimization problems; These solutions are in many cases approximate to the optimal solution. The techniques metaheuristics are useful when solving problems whose model can not be formulated easily or when they have very big. They combine the simplicity of their ideas with their great efficiency to get very good solutions for this type of problems.

Since the last 20 years, several general techniques have been proposed to design these heuristic algorithms. Their ideas are, in many cases, originally based on using models from physics, biology or genetics, among others.

Among the best known are evolutionary algorithms, tabu search, ant colony, particle clouds (Particle swarm optimization) and simulated annealing. It is interest for this research to study evolutionary Algorithms, especially the genetic algorithms, which are based on the mechanism of natural selection of living beings; Therefore they will only be referred to in the next section.

EVOLUTIVE ALGORITHMS OR EVOLUTIONARY COMPUTATION

Recently, evolutionary algorithms have been found very useful for solving multiobjective problems, since they can deal simultaneously with a set of possible solutions (called "population"), and in this way allow to find several individuals in the Pareto optimum set in a single execution, contrary to the case of the classical techniques of mathematical programming (Coello, 2005).

In addition, evolutionary algorithms are less susceptible to some characteristics of the Pareto frontier as concavity, convexity and discontinuity, which are considered as a difficulty for traditional methodologies (Abbass et al. 2001). Another of the advantages that can be identified in this type of algorithm is that given its nature allow you to easily hybridize with other search techniques. Optimization as well as exploiting parallel architectures. In addition to this, they are robust techniques to dynamic changes and their parameters can be self-adapting. As far as their applicability is concerned, it has a fairly wide field and can be traditional techniques, since they are able to solve problems that are still no solution is known (Coello, 2004).

It is possible to identify, within evolutionary algorithms or evolutionary computation, three main paradigms such as Evolutionary Programming, Evolutionary Strategy and Genetic Algorithms (Coello, 2007). The following is a brief description of each one of them.

Evolutionary Programming

It was initially proposed by Fogel et al. (1966). It proposes it as a methodology that rxemplifies the evolution at the level of the species. This empha-behaviors between parents and children.

The algorithm basically consists of randomly generating an initial population, applying mutation and calculate the quality of the children that will define the selection by tournament between them to choose the solutions that will be retained for the next iteration. This algorithm unlike the other two approaches, it does not use the recombination or crossover operator (Santana and Coello, 2006).

Evolutionary Strategy

Peter Bienert, Ingo Rechenberg and Hans-Paul Schwefel (1965) developed a methodology for solving hydrodynamic problems with a high degree of complexity, this methodology was inspired by the mechanism of mutation that occurs in nature (Santana and Coello, 2006).

This approach used a population of 1 parents, from whom only one child was born, who replaced the worst father of the population. Schwefel further proposes two mechanisms of selection, one of them is that the best μ individuals obtained from the union of surviving parents and children and the other is that only μ best children (Santana and Coello, 2006). In this approach there is the recombination operator, which can be either sexual or panmítico. The first of them acts on two individuals randomly selected from the population of parents; And the second is chosen a random father, staying fixed while a second parent is chosen at random for each component of its vectors (Santana and Coello, 2006).

Genetic Algorithms

It was raised by John H. Holland under the name of "genetic reproductive plans" to early 1960s (Holland, 1962). This algorithm uses the crossing, mutation and probabilistic selection, the crossing being the operator (Santana and Coello, 2006).

To implement this algorithm it is necessary to initially represent a random population that constitutes the first possible solutions and from there through a function that evaluate the "quality" of each solution, select "parent individuals" that generate new "child individuals". The latter are obtained from the application of crossing and mutation to individual parents. These children would be closer and closer to be the individuals belonging to the Pareto frontier.

The algorithm requires knowing parameters such as population size, probability of crossing and mutation, maximum number of generations, etc. (Santana and Coello, 2006).

This type of algorithm will be developed in more detail in the following chapter, given which is of interest for this research, since they are recognized as a technique was very successful in solving multiobjective problems of high complexity, and as a very promising technique that can be made some improvements with the in order to increase their effectiveness in solving problems such as combinatorials.

CONCLUSION

There are multiobjective problems that are not easy to solve using the techniques classic optimization, either because the problem has a search space so great that classical techniques would be delayed in finding solutions interesting, because it is not easy to define equations that represent the problem or because these equations are extremely complex, making the classical techniques do not find good solutions.

For these types of problems meta-heuristic techniques such as Evolutionary Algorithms, which have proven to be a very good tool in the solving real problems. Within this type of algorithms belong the genetic algorithms which are recognized as a very good technique in multiobjective problem solving. It is this type of algorithm that is the object of this investigation, so you will learn more about them in the next chapter.

CHAPTER 3

GENETIC ALGORITHMS IN OPTIMIZATION

Ivan Stanimirović

INTRODUCTION

Genetic algorithms have as operators, the selection, the crossing and the mutation. This chapter will describe each of them and some proposals that have developed about its implementation.

Section 3.2 describes how a genetic algorithm works. Describes the selection process and the types of selection they have developed, in the Section 3.4 the types of crossing and finally in section 3.5 describes how mutation operator works.

DESCRIPTION OF A GENETIC ALGORITHM

This algorithm was developed by John H. Holland (1962), which proposes it motivated for solving learning problems of a machine (Coello, 2004).

This is a probabilistic, iterative and directed search method. Initially considered decision variables that could only take values of zero and one, but later versions of the method have been extended to solve problems whose decision variables can take values other than zero and one, representing them in binary code, as shown in Diagram 1.

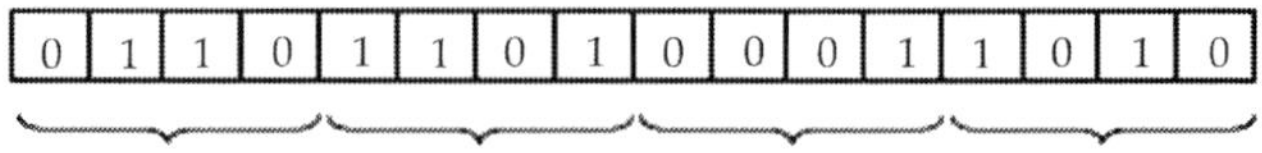

Diagram 1. Representation of a solution in the genetic algorithms taken from Coello (2004).

The entire binary string is called a "chromosome" and represents a solution. To each subsection of a chromosome is called "gene" and constitutes the value that a dcision variable (Coello, 2004).

The coding of the parameters (eg, binary) is known as "genotype" and the decoding of the chromosome, is called "phenotype", ie the values obtained (Binary) to that used by the objective function (Coello, 2004).

The procedure of a traditional genetic algorithm consists of three operations selection, crossing and mutation, which will be described in more detail later. The process starts by randomly generating a set of vectors or strings (Chromosomes) known as individuals, representing a sample of the population total of possible solutions. All these individuals are known as population. Each of these individuals has its own characteristic and is related with the value of the objective function, which is called "fitness", "fitness" or "quality" of the chain, and that from now on will be called fitness (f_i).

The selection operation chooses from the current population (whether initial or not) the alternatives that they have the greatest aptitude for, from these, to create new individuals and thus enhancing their characteristics within the new populations. The creation of new individuals are given through the operations of crossing and mutation. He is a genetic exchange between pairs of individuals (called parents) with very high aptitude value. And the mutation consists of changing some values of the genes, randomly selected from the individuals obtained after the crossing, thus, if the value of the selected gene is one, it is changed to zero and vice versa. The mutation is performed with the order to avoid a premature convergence to local optimities. Individuals products of the crossing and mutation of the parents are called children and replace the parents in the new population or "generation."

SELECTION

This is an important part of genetic algorithms and unlike evolutionary strategies, this is a probabilistic process, that is, individuals with lower fitness value, still have probability of reproducing. There are several techniques that can be used in a genetic algorithm, we will describe some of them, classifying them in proportional selection, selection through tournament and uniform state selection.

Proportional Selection

It was initially proposed by Holland (1975) and proposes to choose individuals according to their aptitude value with respect to the total population. Within this type of selection can mention the following techniques, which will be explained by an example.

Example: We have a population of four individuals, with the following representation binary and the next fitness value.

	Chain	**Fitness**
1	110100	220
2	011010	140
3	111001	315
4	001101	42

Roulette

It was proposed by De Jong (1975) in his doctoral thesis and its use in genetic algorithms has been very common. This technique consists of the following steps:

- Add the aptitude value among all individuals:

$$\sum_{i}^{Tampob} f_i = 220 + 140 + 315 + 42 = 717$$

- Find the probability of selection for each individual:

For individual 1:

$$\frac{f_i}{\sum f_i} = \frac{220}{717} = 0.31$$

For all individuals

Probability of selection

1	220	0.31
2	140	0.19
3	315	0.344
4	42	0.06

- Generate a random number r where r [0,1]

r= 0.6

- Traverse individuals from the population by adding the likelihood of selection to

That the sum is greater than or equal to r

(1) sum = 0.31 <0.6

(2) sum = 0.50 <0.6

(3) sum = 0.94> 0.6 then select individual 3 as parent.

Stochastic Stock

It was proposed by Booker and Brindle. Taking the above example, and normalizing the value of fitness of each individual by F, where F is the average fitness of all the chains, we have Aptitude and Average.

This technique consists of the following steps:

- Determined parents are determined by individuals who have whole parts.

Parents: (1) and (3)

If the population of parents has not yet been completed, it is completed according to the values remaining (difference). To complete the population there are two variants:

No Replacement: Consists of using the remainder to skew a coin and know if a individual is again selected as a father or not.

Shot (1) = 0.23

Shot (2) = 0.78

Shot (3) = 0.76

Shot (4) = 0.23

Proceed to complete the population of required parents. With Replacement: Leftovers are used to dimension the segments of a roulette wheel and then the individuals are chosen in a traditional way, as follows:

		%	
1	0.23	0.115	11.5%
2	0.78	0.39	39%
3	0.76	0.38	38%
4	0.23	0.115	11.5%
	$\sum = 2$	$\sum 1$	

Difference% of Total= % of the Total Difference /$\sum_i$ Difference

Universal Stochastic

It was proposed by Baker with the aim of minimizing the maldistribution of individuals. It consists of the following steps:

- Calculate the normalized fitness, as described in the stochastic surplus.
- Generate a random number between 0 and 1. For example, r = 0.5
- Check if the normalized fitness of the first individual is greater than the number random, if one chooses the individual as a father. Therefore, those individuals that have their normalized fitness value greater than 1, will always be selected. So:

(1) 1.23> 0.5 Individual (1) is selected as parent.

- After selecting an individual as a parent, 1 is added to the random number (r = 1.5) and now this new number will be checked with the fitness value of the individual you are evaluating. So:

(1) 1.23 <1.5 The individual (1) is not re-selected as a parent.

- To evaluate the next individual, the previous fitness value is accumulated with that of the current individual and is compared to the new value of r, for example we have:

(2) 2.01> 1.5 Individual (2) is selected as parent.

- Continue the process until you select the number of parents needed, as follows:

(r = 2.5)

(2) 2.01 <2.5 Individual (2) is not re-selected as parent.

(3) 3.77> 2.5 Individual (3) is selected as parent.

(r = 3.5)

(3) 3.77> 3.5 Individual (3) is selected again as parent.

Therefore, the parents selected in the example would be the individuals: 1, 2, 3, 3

Deterministic Sampling

It is similar to stochastic surplus, but requires a sorting algorithm. Consists in:

- Calculate the normalized attitude, as does stochastic surplus.

	Fitness	Fitness normalized	Whole	Fractions
(1)	220	1.23	1	0.23
(2)	140	0.78	0	0.78
(3)	315	1.76	1	0.76
(4)	42	0.23	0	0.23
	$\sum = 717$	$\sum = 4$	$\sum = 2$	

- Determinedly as the parent, individuals with a part

Individuals with an integer part in their standard fitness, that is, in the example, individuals (1) and (3) are selected, as parents. To complete the number of parents, the population must be ordered according to the decimal parts from highest to lowest, and the top of the list is chosen.

Individual	Fractions Ordered
(2)	0.78
(3)	0.76
()1(	0.23
(4)	0.23

Then the parents who complete the list are (2) and (3).

Selection by Hierarchies

It was proposed by Baker in order to avoid the premature convergence in the techniques of proportional selection. It consists of selecting individuals based on their rank and not based on their aptitude. The algorithm is as follows:

- Sort the population based on their aptitude, with 1 being the least fit.
- Find the evaluation value for each individual where n is the number of individuals, max is a parameter between 1 and 2 and min = 2 - max. Baker recommends that max = 1.1 for individual 1:
- With this value of evaluation the proportional selection is applied to find the parents.

Tournament Selection

This technique was initially proposed by Wetzel and consists of making the selection depending on the direct comparison between individuals. This technique has two approaches:

Deterministic

Continuing with the example:

- Shuffling individuals from the population.

	Fitness	Shuffle
1	220	3
2	140	1

3	315	4
4	42	2

Choose a number of individuals to compete for being parents, usually 2 individuals.

(3) and (1) (4) and (2)

- Compare these individuals based on their aptitude value and choose the individual who

Has greater fitness value as a tournament winner and as the first parent. For him example:

3 and 1: The winner as the first parent is the individual (3) 4 and 2: The winner as the first parent is the individual (2)

- Populate p times to select n parents, where n is the size of the population. in the example, since p is 2, the population must be shuffled twice.

The winners as the second parent are the individuals (3) and (1). Then the parents would be: (3) and (3); (2) and (1)

Probabilistic

It works just like the previous algorithm, only it is decided to choose the individual that has greater fitness value with a given probability r, where $0.5 < r < 1$

Uniform State Selection

This technique was proposed by Whitley, is to replace the μ individuals worse than one generation by the better individuals of the next generation. Then the algorithm is described:

- G is called the original population.
- Select R individuals $(1 < R < G)$ from among those with the highest aptitude value, for example R = 2.
- Execute the crossing and mutation to the R individuals that have been selected and get the children (H).
- Choose the best μ individuals of H, ie the μ individuals with the highest value of aptitude among the children.

- Replace the worst individuals of G by the best individuals of the new generation (H).

CROSSING

This is a process that simulates the crossing between two chromosomes in a system biological. It consists of taking two strings of chromosomes or individuals and exchanging segments of them. This is the most important operator in a genetic algorithm and works as an operator of exploitation, that is, crosses the individuals in order to concentrate your search in the same area. There are three basic crossing techniques:

Crossing a Point

This technique was proposed by Holland in 1975. And it consists in choosing a point of the chromosome chain and combine the segments between two individuals. This type of crossing has a deficiency and is that it considers that the blocks are small, which is contrary to real-life problems where chromosome is quite large and therefore, the blocks that it combines are also, this is why in many cases it often does not present appropriate results(Coello, 2004). Diagram 3 describes the crossing at one point.

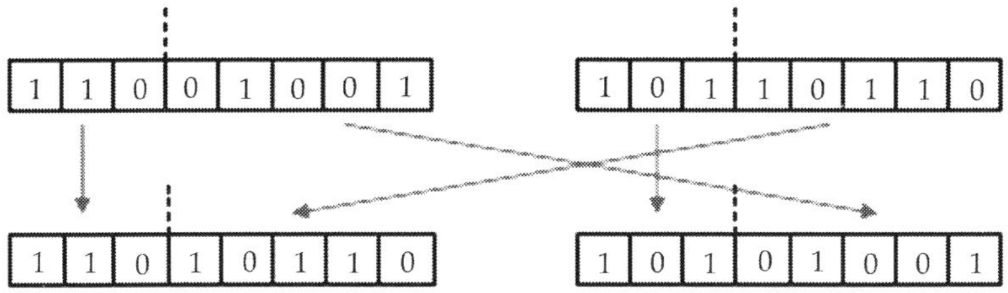

Diagram 3. Representation of Crossing at a Point.

Crossing of Two Points

It was first used by DeJong (1975). The chromosome is split into three segments, of which two of them are crossed and one remains constant. Diagram 4 represents this type of crossing.

Uniform Crossing

This technique was initially proposed by Ackley (1987). It consists of making a cross of n points, only that the points are not known in advance. This is the one that produces the greatest rupture between the three techniques, which should be controlled using a low probability of crossing (P_c<0.5). In Diagram 5 you can see how it works this type of crossing.

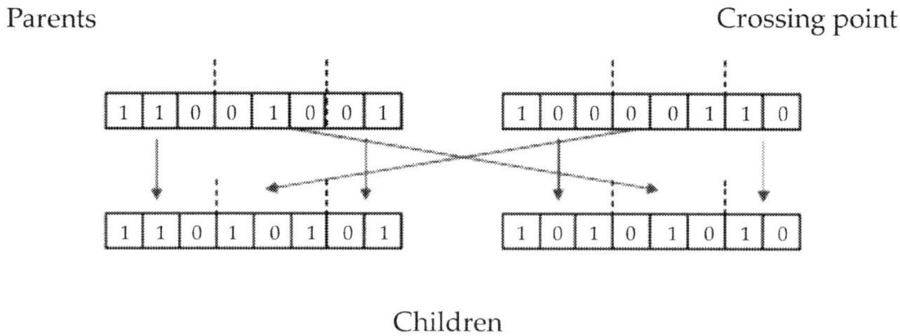

Diagram 4. Representation of Crossing in two points.

MUTATION

In genetic algorithms, this operator is used less frequently in comparison with the crossover operator; Is an exploration operator, since allows the algorithm to explore in other regions, often helping local best.

This operator is carried out with a not very high probability, in practice they are used between 0.001 and 0.01 for a binary representation. However, some researchers suggest the use of high mutation probabilities at the beginning of the process and to the extent that advance the number of generations decrease the probability of mutation. Others researchers suggest that the probability of mutation is:

$$P_m = \frac{1}{TamCadena}$$

used as minimum (Coello, 2004).

For problems where the decision variables take values of 0 and 1, the operator ofmutation functions as described in the following algorithm and is represented in Diagram 6.

- The decision-maker must determine the probability of mutation, for example Pm = 0.009 P_m
- Find a random number between 0 and 1 for each gene, for example R_1 =0.2 1
- Determine whether or not to change the value of the variable. If $R_1 < P_m$ mutates the variable, otherwise it would not mutate.
- The process is repeated for each variable.

Currently, there are diverse techniques for the selection, crossing and mutation of a genetic algorithm, both of which are used to solve problems monobjective as multiobjective. Determining which one to use depends on who is making use of this methodology.

Having a clear understanding of how a genetic algorithm works, the next chapter will describe the use of this methodology to solve multiobjective problems, for this purpose they are described the main algorithms that have made history in the field of genetic algorithms multiobjective.

CHAPTER 4

APPLYING GENETIC ALGORITHMS TO RESOLVE MULTIBJECTIVE OPTIMIZATION PROBLEMS

Ivan Stanimirović

INTRODUCTION

In the last decades, the use of genetic algorithms has increased. Solve multiobjective problems, thanks to its strength of being able to work simultaneously with a set of possible solutions that are increasingly efficient and satisfactory results.

Due to the importance of genetic algorithms, some researchers have proposed different improvements in selection, cross-breeding and mutation in order to solutions closer to the Pareto real border and better length of it. Section 4.2 will briefly describe some of these algorithms that have marked the history during the last years in the literature of the algorithms multiobjective and some observations will be made on these.

Since it is important that the decision maker always involves his / her preferences when it comes to solve multiobjective problems, genetic algorithms have also been adapted to that comply with this principle, some of these algorithms are mentioned in section 4.4 and some of its weaknesses and strengths in section 4.5, finally in the Section 4.6

mention the characteristics that would be taken into account in a new methodology.

MULTI-OBJECTIVE GENETIC ALGORITHMS

Over the years, authors have proposed various genetic algorithms for multiobjective problems always looking for a Pareto front better delineated and with better solutions. At present, it has added to this objective the efficiency of algorithm, that is to say that the solution that throws up is in the shortest possible time.

The first time we mention the use of genetic algorithms to solve multiobjective problems is in 1967, by the researcher Rosenberg, who mentions it in his doctoral thesis to solve a biobjective problem of chemistry, but does not implement the algorithm only exposes the problem in a modified way, such that it since then different methodologies have been developed that aim to obtain very good multiobjective problem solutions. Some of them will be mentioned in the this chapter.

VEGA

It is an approach posed by David Schaffer (1985), and is considered as the first methodology that is actually proposed to solve a multiobjective problem. Approach was called Vector Evaluated Genetic Algorithm (VEGA) which is an algorithm gene with a modified selection mechanism.

This algorithm consists of subdividing the population into so many objectives and each subpopulation evaluates it with a specific objective, that is, it selects a future subpopulation based on a single objective omitting the rest, so does the rest of subpopulations.

Then he mixes all the individuals, trying to get an individual who was the best in a sub-population is re-combined with another that was better in another subpopulation and can get a better population. This algorithm presents problems because the selection can remove an unmanaged solution that can be a very good solution to combine. It works well in problems where there is no conflict between objectives and where they use simple functions with a single variable (Coello, 2005).

This method was criticized by the researcher David Golberg in 1989, who from this critique proposes the use of non-dominated categories to move the population toward pareto frontier, that is, to find solutions that dominate the rest of the population. It proposes to assign an aptitude value to the alternatives based on Pareto optimality, giving the highest aptitude value to those who are not dominated, and the others are assigned proportionally. In addition, it suggests a type of technique (called a niche) for the ability to distribute individuals along the border of pareto. Although Golberg did not propose a specific algorithm, it did induce work that developed later had this same approach (Coello, 2005).

MOGA

In 1993 the researchers Fonseca and Fleming developed an algorithm called Multiobjective Optimization Genetic Algorithm (MOGA). Its contribution to the traditional algorithm (Fonseca and Fleming, 1993) is in the calculation of the aptitude for each individual, which depends on the number of individuals who dominate it. Each alternative has a classification as follows:

$rank(x_i, t) = 1+p^{i(t)}$

where i x is the alternative i to who is being calculated the value of the aptitude in the generation t, and $p_i^{(t)}$ is the number of individuals that dominate i in that same generation. Thus, assuming that a maximization problem is optimized, the non-dominated individuals ($p_i^{(t)} = 0$) will have a rating of 1 and the greater the number of individuals dominating x_i, highest is the rank value for x_i.

Then the population is organized according to the rank of each individual and is assigned the aptitude interpolating from the best order (rank = 1) to the worst (rank = n), according to one function that is usually linear. Finally the average ability of individuals with the same classification, then each belonging to the same order could be selected with the same probability.

This was one of the most effective algorithms in its time, comparing it with VEGA, NSGA and NPGA (Coello, 2005). However, Goldberg and Deb point out that the type of fitness proposed by Fonseca and Fleming, it is possible that the algorithm may have a premature convergence (Coello, 1999).

NPGA

In 1994 researchers Horn, Nafpliotis and Goldberg (1994) developed an algorithm called Niched Pareto Genetic Algorithm (NPGA), which incorporates the concepts of Pareto dominance and niche in the selection operator. The concept of niche is introduced in order to exert pressure to extend the entire population throughout the the Pareto border and is used in the tournament to determine the winner when there is a Tie by dominance. So:

Two potential candidates for breeding are randomly selected from the population, just as a set of individuals is selected for comparison (usually it is the 10% of the total population). Each candidate competes with each individual in the set, winning the tournament who is not dominated by any of them. If the two candidates or none of them are dominated by any individual in the set, then the winner according to the niche in which it is. In this case the candidate who wins belong to the niche that contains fewer individuals, as shown in the following (Horn et al., 1994):

In this selection would win Candidate 2. And the radius of the niche is set by the user as the minimum separation expected between the final solutions. This algorithm, although simple, is more efficient than the MOGA, but slightly less effective than this one (Coello, 2002).

If you want to investigate more about this method and how the niche is calculated, you can investigating (Horn et al.1994).

NSGA

Researchers Deb and Srinivas (1994) created the Nondominated Sorting Genetic Algorithm (NSGA). These take the idea proposed by Goldberg, which consists in classifying the individuals by fronts depending on the dominance.

This differs from a simple genetic algorithm by the way it works with the selection operator. Thus, to select the parents take into account the following procedure: to find in the present population who are not dominated and to constitute the first front. These individuals are assigned the same high dummy proficiency value.

In order to maintain the diversity of the population, the individuals at the front are distribution operator (Sharing) who tries to distribute the points on the border, which it does by dividing the dummy fitness value of an individual by an amount proportional to the number of individuals around them. As follows:

$$f_i = \frac{f_i^*}{m_i}$$ where $f_i^=$ is the fitness value dummy y

$$m_i = \sum_{j}^{pob} Sh(d_{ij})$$

This last amount is calculated according to a Sharing operator, like this:

$$Sh(d_{ij}) = \begin{cases} \left(1 - \dfrac{d_{ij}}{\sigma_{Share}}\right)^2 & d_{ij} \leq \sigma_{Share} \\ 0 & d_{ij} > \sigma_{Share} \end{cases}$$

where Ω_{share} s is the radius of the niche set by the user and represents the separation minimum expected between the final solutions, and d_{ij} is the distance of the individual i to individual j.

After distributing it, these non-dominated individuals are temporarily ignored to find the next non-dominated border among the remaining individuals. To the latter boundary is assigned a new dummy fitness value less than the smallest distributed from the previous frontier, this is done until classifying the entire population in several borders.

Once the classification is finished, the selection is done taking into account the value of aptitude dummy this algorithm was not very efficient because the Pareto boundary must be repaired many times, therefore it turned out to be a very slow, ineffective algorithm and implied high costs in its implementation (Coello, 2002).

SPEA

The researchers Zitzler and Thiele (1999), created an algorithm called Strength Pareto Evolutionary Algorithm (SPEA) which uses an external file to retain the solutions non-dominated domains previously encountered; In each generation is copied in this archive the non-dominated solutions and delete those that happen to be mastered. If the number of individuals not mastered in the external file exceeds the given number allowed N 'then, individuals are reduced by cluster. SPEA does it with the average linkage method. This method consists of the next:

1. Initialize the cluster cluster C, where each cluster is made up of an individual of the uncontrolled external set $C = \bigcup_i \{i\}$
2. If $C| \leq N'$ then go to step 5, but continue with step 3.
3. Calculate the distance of all possible cluster pairs. The distance between two cluster, for example 1 and 2, is given by the average distance between all possible pairs of individuals belonging to the two clusters as follows:

$$d = \frac{1}{|c_1| \cdot |c_2|} \cdot \sum_{locl, joc2} \|d_{ij}\|$$

   where $Id_{ij}I$ represents the Euclidean distance between individuals i and j
4. Choose the two clusters with the minimum distance and merge it into a single cluster. Continue with step 2.
5. Choose one representative individual per cluster. SPEA recommends the individual centroid, that is, the individual who has the lowest average distance to all individuals in the cluster.

For each individual belonging to this external file is computed a value of force is (strength) which is proportional to the number of solutions to which each individual dominates, like this:

$$s_i = \frac{n}{N+1}$$

where n is the number of individuals in the population that are dominated by i and N is the population size. The fitness for each individual i belonging to this file external is the same value of force $f_i = s_i$.

Each individual j belonging to the current population is calculated their aptitude depending the force values assigned to the individuals in the external file that dominate it. So:

$$f_i = 1 + \sum_{l,lo-j} s_i$$

The value of one is added to ensure that the fitness of the individuals in the file be better than the rest of the population, taking into account that to the extent that the value of fitness is less, the individual will be more likely to reproduce.

The way in which aptitude is posed seeks to be preferred to the individuals closest to the border of Pareto and that less individuals have around it.

After the assignment of the aptitude one can proceed to the selection, which is done by binary tournament, choosing the candidates to be parents since the union of the file with the rest of the population. With these parents the children are created after applying the two operators: crossover and mutation, as presented in an algorithm genetic simple.

This algorithm yielded results well above the other existing techniques to introduce the new concept of elitism, which led to its popularity in the area of multiobjective genetic algorithms.

NSGA-II

In the year 2000 researchers Deb, Pratap, Agarwal and Meyarivan developed a algorithm called Nondominated Sorting Genetic Algorithm-II (NSGA-II) described in (Deb et al. 2002) which uses a rapid procedure to organize the population by not dominance, an approach to preserve elitism and an operator, different from the niche, to dispersing individuals at the Pareto border.

Initially, they organize the entire random initial population into non-dominance boundaries, the first frontier being the individuals that are not dominated within the population. In order to organize the rest of the population, it forgets organized borders and evaluates non-dominated individuals by organizing them on a new frontier.

It assigns the aptitude to each individual depending on the border to which it belongs, assigning as aptitude value 1 to those belonging to the first frontier, 2 to the second, and so to all individuals. It should be noted that this algorithm gives priority to reproduce to the individuals that have less value of aptitude.

Then they select the parents by binary tournament and through the operators of crossing and mutation creates new children. At this point you use the "mas" selection, which consists of the children joining the parents forming a population t R which would be of size 2N. Therefore, this population should be reduced according to the criterion of non-dominance in order to obtain the new population of size N. This last population will be the one that is used for the selection, crossing and mutation to create a new generation.

The procedure for reducing the 2N population is to classify the population into several frontiers and to complete the new population with the best borders value. At the last accepted border, the solutions are selected according to a operator called crowded, which defines the maximum perimeter of a contained box between two neighboring solutions. To use this operator, you must first find the density estimation metric for each individual.

To calculate this, the population according to the value of an objective that each individual takes; This is done for each objective. Individuals who are at the extremes, that is, that have the lowest value in that objective and the one with the highest value, will have a value of distance of infinity. And for the rest of the individuals i the distance is calculated through the following equation:

$$Dist_l = Dist_l + \frac{f_{i+1,j} - f_{i-1,j}}{f_j^{\max} - f_j^{\min}} \quad \forall j$$

where f_{ij}, is the value of the objective j of the individual i+1 and f^{max} ,f^{min}, is the maximum value and target minimum j.

When this metric is found, the Crowded operator can be used, which guides the selection in several states of the algorithm towards a uniform distribution of individuals along the optimal Pareto frontier. With this operator we define that i is preferred to j, provided that i has a better classification in terms of frontier and in the case of that i and j have the same classification, then i is preferred to j when i has a greater value in the metric of the density estimation.

It should be noted that this same operator is used in the selection of parents, that is, in the event that the candidates in the tournament are tied for their value of aptitude, then it is chosen who has a greater value in the metric of the estimation of the density. At present, this and SPEA2 are perhaps the best algorithms, since they find solutions very close to the Pareto real border, so they are taken as reference algorithms for the scientific community and therefore for this research.

SPEA2

In the year 2001 Zitzler, Laumanns and Thiele make corrections to the algorithm SPEA, to the called SPEA2 (Zitzler et al., 2001), and the latter differs from the former by the fitness assignment, parent selection, truncation operator and setting the size of the external file for all its generations.

Aptitude Allocation

For the assignment of the aptitude, unlike the SPEA, takes into account both the dominated and dominated individuals and the density of each person. The strength value of the individual i, i S is given by:

$$S_i = \left|\{j \mid j \in P_t + \overline{P}_t \wedge i > j\}\right|$$

where j is equivalent to the number of solutions that i dominates, P_t and $\overline{P}_t$ are the individuals of the population and individuals belonging to the external file in generation t respectively. Taking into account this new definition of force value, the value of a raw fitness (regardless of density) for each individual i R which is given by the following equation:

$$D_i = \frac{1}{\sigma_i^k + 2}$$

It should be noted that to the extent that an individual has an aptitude value small will be preferred to one having a higher value. Another component of the fitness of each individual is the measure of density i D, which propose as an adaptation to the nearest neighbor method.

To find this value, calculate the distance from i to all individuals, both from the population and from the external file and then these distances are organized in a list in ascending order, and the value of the K - element as σ_i^k where k = . Where N is the size of the population and N is the size of the external file. In this way the fitness value for the individual i . It is given by

Individuals From The External File

All individuals are initially copied dominated by both the population and the external archive of the previous generation, external file of the current generation, one way to identify them is to select those individuals that have a fitness value less than one.

For SPEA 2 the size of the external file is always constant over time, for both, in the case in which these individuals do not add up the amount given by the user N, must complete the file with the best dominated individuals, that is, with those who have lower fitness value.

And if the number of non-dominated numbers found is greater than N, then the individuals needed until the file size is N. The way in which the individual to remove is to find the one that has the minimum distance to other individuals and this will be eliminated, in case there are several individuals with the same minimum distance, is chosen who has the second lowest minimum distance.

Parent Selection

Selection is by binary tournament with replacement only the individuals from the external file participating. With the proposed improvements to this algorithm, it effectively achieves better than the SPEA, and although it has not had the same popularity as the latter, been preferred along with NSGA-II for the optimization of real problems.

cNSGA-II

Deb and Goel (2001) propose to control elitism in NSGA II, since it considers that elitism in multiobjective genetic algorithms leads to prematurely converging to a set of sub-optimal solutions, therefore propose to control it and in turn maintain a balance between exploration and exploitation.

The control consists in restricting the number of individuals that are accepted from each border to be part of the next generation. The number of individuals who are accepted depends on the following distribution:

$$n_i = N\frac{1-r}{1-r^k}r^{i-1}$$

Where n_i is the maximum allowed number of the border i, N is the size of the population, K is the number of frontiers found and r <1 is the rate of reduction, given by the user.

In the case that n_i is greater than the number of individuals that have border i, then is selected by tournament the n_i individuals of the border and in the case that i n is smaller, then all of that border are accepted and the missing ones are chosen from the next, so the number of individuals on the next frontier would increase by that amount.

This would achieve greater diversity in the population and the possibility of finding a better distributed border.

The above methodologies aim to find the entire Pareto frontier and however, in order to solve real-world problems, it is necessary to have a single solution.

Choose this solution among a fairly large group of equally good alternatives is somewhat difficult, so we continue with a discrete decision problem. Deb and Chaudhuri (2005) mention that in addition to the problem of finding a set of optimal solutions is the scientific selection among them a solution particular.

In order to obtain a single solution or a small group of use of the preferences of the decision maker in this type of methodologies. The next section describe some proposals of how to incorporate the preferences in these algorithms.

INCORPORATION OF PREFERENCES IN GENETIC ALGORITHMS MULTIBJECTIVE

Incorporating preferences into methodologies that solve multiobjective problems is great help to increase the selection, since it emphasizes solutions that are relevant to the decision maker. In addition, it can also help address better the search for those problems that involve many objectives and that is difficult to do so based only on the Pareto dominance criterion (Rachmawati and Srinivasan, 2006).

However, incorporating the preferences has associated a great difficulty, due to the uncertainty (due to the previous lack of knowledge of the problem) in the preferences of the decision maker that are always very diffuse. In turn, the complexity of the makes the mathematical formulation of preferences very difficult (Rachmawati and Srinivasan, 2006).

Who developed the first scheme that incorporated the preferences of the decision-maker in the Evolutionary Algorithms Multiobjective was the researcher Masahiro Tanaka in 1992, and from this point on, this topic was considered as a very important topic since it was very much in the solution of real life problems, since they demanded to obtain only few points or only one and not the whole border, however, was a subject in which few researchers showed interest (Coello, 2005).

Since then, they have developed several proposals, some of them we will mention below making use of the classification made by Veldhuizen and Lamont (2000), who propose to classify the multiobjective genetic algorithms in a priori, interactive and a posteriori, according to the moment in which they are incorporated preferences.

A priori Algorithms

Algorithms a priori formulate preferences prior to optimization and usually are represented by parameters that help to formulate a single function objective through aggregative objectives. This type of algorithm does not allow changes in parameters during the search (Rachmawati and Srinivasan, 2006).

This approach is very simple and efficient when you have a good representation of the decision maker's preferences and do not present difficulties at the time of compute them, which is difficult to do. In addition, it requires a very good knowledge of the possible results of the model (Rachmawati and Srinivasan, 2006).

In 1998, Voget and Kolonko combine fuzzy logic with evolutionary algorithms, with a fuzzy controller that automatically regulates the selection process of an evolutionary algorithm, through a set of predefined goals by the user that define the desirable behavior of the population

(Coello, 2000). Since these goals are provided at the beginning of the process, this algorithm is categorized as a priori.

In 2001, Branke J, Kaubler T, Schmeck H, propose to involve preferences through the representation of sacrifices between objectives. A priori specify two sets of weights that correspond to the maximum and minimum sacrifice of each one of the objectives with respect to others, which is willing to assume the decision maker. Then use each set of weights to find an aptitude function by a weighted sum and given that they are two groups of weights, we would have two fitness functions to evaluate each alternative. In this way, it will dominate the alternative that when evaluating it has greater value (in the case of maximization) in at least one of these two functions. This methodology has disadvantages such as not being able to model the dependence between objectives or the difficulty of specifying minimum and maximum sacrifices when more than two objectives (Rachmawati and Srinivasan, 2006). In 2002, Cvetkovic and Parmee propose a methodology that models preferences through the relative importance of the objectives with the help of linguistic variables. Constructs an array of fuzzy preferences to calculate the parameters that are then used with the weighted sum method. The weakness of this methodology is that it is very sensitive to the required parameters (Rachmawati and Srinivasan, 2006).

In the same year (2002), in order to point to the weakness presented by the methodology proposed by Cvetkovic and Parmee, Jin and Sendhofff use intervals of weights (Rachmawati and Srinivasan, 2006), representing preferences through the importance relative between objectives using linguistic variables such as: very important, moderately important or unimportant.

These variables are transformed into values using methodologies such as random weighted sum and sum dynamic, which are described in Jin and Sendhoff (2002), taking into account that each belongs to a given interval previously. Once the matrix is converted in real numbers, these inputs help to calculate the parameters that will be used in the weighted sum (Jin and Sendhoff, 2002). This methodology presents weaknesses as not save computing time, do not converge very well on concave fronts and not show dependence between objectives, a characteristic that is common in the real life (Rachmawati and Srinivasan, 2006).

In 2003, Tan, Khor, Lee and Sathikannan, represent the preferences specifying multiple goals through conjunctions such as AND and OR and classifying objectives by priorities. In this way it selects the non-dominated Pareto front alternatives. This methodology has as a disadvantage the computational complexity and the notions of compensation or any other relationship between objectives (Rachmawati and Srinivasan, 2006).

In the same year (2003) Filomeno, Bersini and Bouillard, propose a methodology combining the evolutionary algorithms with PROMETHEE II (Coelho et al., 2003) where take into account both the preferences of the decision-maker and the constraints comply. In this methodology constraints are taken as a more objective function, which relates the satisfaction of meeting the restrictions. In the selection, the two new children with the two parents who created them using PROMETHEE II, and choose the two best to be part of the new generation. The weights with which initiate this methodology are given by the decision maker, but as the process progresses these are self-adjusting according to the compliance of the restrictions. It must be taken in that in this methodology the weights are self-adjusting according to the restrictions, but are not modified by taking into account the changes in the preferences initially given by the decision maker.

Cvetkovic and Coello in 2004 mention in one of their articles (Cvetkovic and Coello, 2004) A co-evolutionary multiobjective genetic algorithm that consists of optimizing m objectives using m algorithms in parallel, such that each of them optimizes a target and during the run process a penalty function is used progressively, according to a distance in the genotypic space, to guide the solution to a common area. In the first iterations is a bit flexible when accepting solutions far from the standards established and in the last generations a variation of more than 10% of the range of variables. The most important objective for the decision-maker is less penalized that the least important and thus the solution obtain good values for the objectives of greater interest.

In 2006, Hisao Ishibuchi, Yusuke Nojima, Kaname Narukawa, and Tsutomu Doi, Decision-makers' preferences as a way of improving the solutions while preserving the diversity between them. They propose to make a hybrid between two approaches to multiobjective optimization: Multicriteria Decision Making (MCDM) and Multiobjective Evolutionary Optimization (EMO). The first in order to make a efficient search and the second in order to give the decision maker a set of solutions and not a single solution. The hybrid consists of involving a priori preferences in a scalable function that will be used to select the parents of the new generation. The way in which they are involved will depend on the decision-maker's criteria, that is, when relative importance between the objectives, then you will use the preferences in a weighted sum fitness test, or when it is required to meet a minimum level of goals, will reflect preferences in a role that penalizes parental selection when the individual does not meet the required minimum of the target, or if he has a solution that represents the preference of the decision maker, then the selection of the parents is given by the distance to this reference solution (Ishibuchi et al., 2006).

This, like all methods that allows the decision maker to define his preferences of form a priori have difficulties, since the decision maker would have to have them well defined for the problem you are solving and this in most cases is not possible. In addition, the parameters representing the preferences, sometimes could be very optimistic, out of the region feasible or sometimes so unrealistic which would be lost in obtaining better solutions.

Interactive Algorithms

In the interactive approach preferences are formulated prior to optimization, but are modified by the decision maker during the search process, allowing him to influence the address. These algorithms are the least common in the literature.

This approach takes into account the learning that comes from the problem with the information that is revealed during the search, which makes the decision-maker judgment to give value to the parameters (Rachmawati and Srinivasan, 2006).

One of the first algorithms proposed under this approach was presented in 1993 by Fonseca and Fleming and mentioned in Coello (2000). They propose a methodology based on MOGA by adding an interactive achievement goal that allows to be modified by the decision maker as he becomes more aware of the problem. Recommend the construction of an expert system based on the knowledge about the preferences of the makers to automate achievement goals.

Then, in 1998 Fonseca and Fleming propose to prioritize objectives and compare solutions starting with the highest priority objective. This methodology presents weaknesses as high computational complexity and does not allow the modeling of compensation or any another relationship between the objectives (Rachmawati and Srinivasan, 2006).

Deb and Chaudhuri (2005), suggest a tool that allows the decision maker to find either the whole border of Pareto or concentrate only in a specific region. It also gives you the decision maker to obtain "knee" solutions (solutions where the decision-maker does not want move, as it involves a high sacrifice of one of its objectives to increase one unity of the other) or obtain a robust border.

For this, initially the tool makes use of an Evolutionary Algorithm like the NSGAII to find an initial boundary. This border is improved implemented several concepts; For example, to improve the ends, the tool has the use of genetic algorithms for a single objective (corresponding to the end desired improve) and to improve the intermediate sector proposes to apply a search algorithm local for some border solutions found with the NSGA-II. These solutions can be automatically selected by taking the ones with the lowest density or selected according to the preferences of the decision maker from certain methodologies such as sum weighted, utility function, among others, which will depend on the decision maker.

If the boundary improves, the algorithm continues to work with the final frontier, contrary to that found with NSGA-II. Finally try to improve it again using the classical methods of e-constraint (Deb, 2001), where the decision maker must define the parameters.

The decision-maker decides whether to stay with this border, to concentrate on a particular region or stay with a specific point. If you want to concentrate on a region, improve again is as described above. This procedure repeats as many times as necessary until the desired point or fraction of the border is obtained.

Within the interactive algorithms we could mention other methodologies that algorithm as it runs, such as presented in Cvetkovic and Coello (2004) where they mention that in some occasions, preferences should not necessarily be given by the decision-maker or group of which can be determined mechanically using agents, as is the case in engineering designs. As the algorithm runs the preferences are adjusted according to whether or not certain restrictions are met. However, in this way the preferences would not be represented since it would seek to modify the parameters such that a border can be obtained closer to the real or better distributed, but would not focus on a fraction of interest given by the decision maker.

An algorithm under the interactive approach is very attractive, since it allows the decision maker to change their opinion as to their preferences as they come to know the possible results, however, require a greater mental effort for the decision to adjust their preferences, it is therefore algorithms in this approach are practical in defining how represent the preferences of the decision maker.

A Posteriori Algorithms

In the a posteriori approach the preferences are only involved until the algorithm finds a set of non-dominated solutions. Under this approach the decision-maker avoids problems of computation, however appear other problems as the cost computation, the difficulty of exploring the optimal Pareto set to select a single solution when you have a problem that considers many objectives, and the little progress in the search for the optimum when they have a search space large, as is the case with problems with many objectives (Rachmawati and Srinivasan, 2006). Added to this, despite the effort to find the whole front, it is possible that finally offers the decision maker a set of alternatives that does not fully comply with their preferences, compared to the actual border.

The algorithms categorized in this approach are the evolutionary algorithms described in Section 4.2, which find a possible Pareto border and from there, they give the decision maker the freedom to choose the solution that would implement. To decide what solution implement, the decision maker can make use of concepts such as outranking, interchange between objectives, utility function, etc.

Massebeuf et al., For example, in 1999 put forward a methodology that uses the method VEGA to find the Pareto curve and then classify the alternatives found in the curve by the outranking criterion, finding an index of preferences of one candidate solution to another. This methodology presents problems, since the as you increase the number of targets you would need to specify many parameters, requiring a lot of effort for its configuration. In addition you would get points that are not necessary for the decision maker (Rachmawati and Srinivasan, 2006).

Under this approach the decision maker is limited to selecting the alternative from the non-dominated solutions obtained with the multiobjective genetic algorithm that is being using, therefore, in the case of not obtaining the entire Pareto border, the decision maker must choose among which the methodology can offer, even if there may be a solution that best suits your preferences.

WEAKNESSES AND STRENGTHS IN METHODOLOGIES WITH PREFERENCES

The multi-objective analysis is a vision of much more optimization problems realistic and complete, however, the complexities added by the conception, formulation and resolution of this type of problem has not applied.

Some of these complexities are that there is no "optimal" solution under the concept classic optimization, ie there is no single solution that can not be overcome in the feasible region, but there are infinite efficient or optimal-Paretian solutions. However, the decision-maker needs to choose only one of them in order to implement it, for both must resort to identifying their true preferences; Which would be a task difficult and sometimes even impossible.

This is the case of solutions obtained by algorithms such as NPGA, NSGA, SPEA, NSGAII, SPEA 2, etc. Seen in section 4.2, who have demonstrated great efficiency in time to find the Pareto frontier in complex multiobjective problems such as such as high non-linearity, high dimensionality, combinatorial problems, etc. However, these methodologies suffer from some shortcomings that will be following section.

Faults Found in the Literature

The methodologies seen in section 4.2 and that when incorporating the preferences are used in the a posteriori algorithms present some shortcomings, one of these is that problems like the backpack find only a discrete set of the "elbow" of the Pareto border, without considering that the solution that meets the true preferences of the decision maker is outside that elbow.

Hisao Ishibuchi and other researchers (2006) argue that the NSGA II, being one of the algorithms now provides good candidate solutions when is solving the problem of the backpack, but when the decision maker is interested in the central area of the Pareto border, but under the assumption that the decision maker wishes to obtain a very good value of one of the objectives, this algorithm can not provide a good candidate solution.

Also, in the case of a posteriori algorithms the decision process continues incomplete, because it would still be necessary to implement an additional analysis methodology multicriterio that allows to incorporate the preferences of the decision maker to obtain the solution last to implement.

This last process, in addition to increasing the computational cost, creates the use another methodology (including maybe other software, other learning, etc.), invest more time to get the answer to the problem, create the possibility that the true "preferred solution" is not among the discrete set that throws the algorithm genetic, etc., does not free the decision maker from the task of having to express their preferences. In addition to this, if the discrete set of possible solutions is very large, the decisions is very complex, and if, on the contrary, it is small, this can be very simplistic.

Although it is difficult for the decision-maker to be clear about "his true preferences" at the end of the process of a genetic algorithm, it is even more at the beginning of the process, as is the case of algorithms a priori, since the decision maker does not know their possibilities, the values maximum or minimum levels that could be achieved in the different objectives, feasible and efficient between objectives or other information that allows you to have an idea about the effects that your preferences might have on finding a solution.

Deb and Chaudhuri (2005) state that the existing methods demand an amount of the problem by the decision-maker, without providing the decision maker with any idea of the nature of the result of the optimal Pareto frontier, causing it to have Guess what is happening.

As will be mentioned in the following section, it is possible that these deficiencies may be solved with an interactive algorithm, however the latter may require the decision maker a high degree of mental effort in deciding on their preferences, for both this type of methodology also require care when defining how to involve the decision maker.

Strengths Found in the Literature

In the literature we also find, although very few, a number of algorithms interactions that seek to solve these difficulties.

By incorporating the preferences of the decision maker into the same genetic process, one could obtain solution or a small set of optimal solutions that are easy to implement. Analyze, with an equal or sometimes less computational cost, because the search space would be reduced. This type of methodology gives the decision maker the possibility of not defining their preferences definitive at the beginning of the process, but as the process progresses, the decision-maker is obtaining new information that allows you to learn, analyze the possibilities and go clarifying and defining their preferences. This type of information may be related with the best solutions found so far, the maximum and minimum value that take the objectives during the process, trends, possible exchanges, etc.

This is because preferences can not be independent of the problem itself and the real possibilities, since the decision-maker, in general, before objective against another, you want to know what level of benefits will be guaranteed. Therefore in a single process and with the same computational time the decision maker can have the final solution to be implemented, or at least a small set of them, which will be much easier to analyze. Since the search will be guided by the decision maker, this small set will involve the alternatives of interest to him, which will not necessarily form the central zone of the Pareto border, but will be located according to the preferences of the decision maker.

CONCLUSION

Knowing the advantages that an interactive methodology can provide, but taking into account that it is important how the decision-maker engages in it, in this research intends to use this approach to raise a new methodology that gives the decision maker a set of solutions, not so great as that is not clear when choosing the alternative to implement, nor so small so that you feel insecure when making a decision.

This new methodology must provide security to the decision-maker when preferences, ie that it can do so in a practical and re-evaluate your preferences whenever you want.

It is expected that the decision maker in facilitating his preferences guide the algorithm search to the region of their interest, and at the same time limit the search space, such that the computational cost that was used to search in a wider area, used now to search in a smaller area, so this additional effort can be used to improve the quality (effectiveness) of the border found.

Initially, the main concern of researchers in this field was to improve the so that they could offer the decision maker multiple solutions that are close to the border, however, they forgot the importance that it has for the decision maker know a good solution but according to your preferences regarding the problem that is solving.

In recent years, the issue of preferences has become increasingly important and incorporated into multiobjective genetic algorithms before (a priori), during (interactive) or after (a posteriori) to obtain the Pareto frontier, algorithms that involve your preferences before you start the process because of your simplicity.

However, it must be emphasized that the interactive algorithms are quite promising, because they allow the decision maker to adjust their preferences according to what is learning from the problem and concentrate the search only in their region of interest, which could lead to solutions closer to the Pareto frontier in the same number of iterations, since there is a smaller search space; Besides, problems such as backpacks, they could find solutions that were not obtain with an algorithm a posteriori.

In the next chapter, two new methodologies will be proposed, one that will Pareto and an interactive one that makes use of the benefits of this type of methodologies to obtain only a portion of the Pareto frontier in the interests of the decision maker, and perhaps closer to the real border than if a very good genetic algorithm were used multiobjective without preferences.

CHAPTER 5

TWO METHODOLOGIES FOR SOLVING MULTIOBJECTIVE PROBLEMS USING GENETIC ALGORITHMS

Ivan Stanimirović

This chapter describes two methodologies for solving multiobjective problem, one of them is described in section 5.1 and consists of making use of the strengths of NSGA II and SPEA2 to propose a new methodology to find a border of Pareto near the real one. While the other methodology, described in section 5.2, is to incorporate the preferences of the decision maker in an interactive, such that it can obtain a portion of the Pareto border that is of interest and much easier to analyze.

MULTI-OBJECTIVE ELITISTIC GENETIC ALGORITHM – MOEGA

Within the scientific community of multiobjective genetic algorithms, the NSGA-II and the SPEA2 are recognized as the reference algorithms, since they obtain solutions very close to the Pareto real border and well distributed along this his efficiency is due to novel concepts that manage each one of them.

The NSGA-II, for example, gathers the parents with the children and chooses the best ones to conform the new generation, this makes the algorithm in the same number of iterations get results closer to the real frontier than join them. In the case of SPEA2, this improves its effectiveness by creating an external file that allows you to save and with the best individuals of each generation, who also compete with the time to select the parents to be reproduced.

Given that within the genetic algorithms are two interesting concepts, in this research intends to make use of them to create a new methodology multiobjective and verify if the results could be closer to the real border. The proposed algorithm will be called Multiobjective Elitistic Genetic Algorithm (MOEGA) and its main characteristics are described below:

Handling of Restrictions

MOEGA works with feasible individuals, therefore each infeasible individual found within of a generation, as well as the infeasible children generated from the parents, should be repaired with a 100% probability. The repair is done based on the "Greedy Repair" technique, used by Eckart Zitzler (1999) in his algorithm SPEA multiobjective.

The repair consists of removing one by one the items of the solution until it comply with all restrictions. In the case of the backpack problem, which is the test problem in which the algorithm will be evaluated and which will be described later in the Section 6.2, the first item to be deleted is determined by the maximum gain / weight ratio i q la which is found by the following equation:

$$q_i = \max_{j-1}^{m} \left\{ \frac{c_{ij}}{a_{ij}} \right\}$$

Where c_{ij} represents the gain given by the decision variable i to the objective function j and a_{ij} represents the weight (technological coefficient) that has the decision variable i in the restriction j. It should be noted that for the problem of the backpack, the number of the same number of constraints, so for other you must use another strategy to remove the items.

The item with lower i q is the first to be eliminated, thus it is obtained that the variables of decision that give less gain to the objective function with respect to its weight, are the first to be removed, until an individual can be obtained that complies with the restrictions while the profits of the latter are sacrificed as little as possible.

Dominance

The concept of dominance that is handled in MOEGA is the classic concept, where in a problem of maximization, an individual i dominates an individual j, if individual i is equal in all objectives the individual j and is greater than the individual j in at least one of the objectives.

Distance Factor

Studying the behavior of a genetic algorithm, it was observed that these tend to find all their individuals very close to each other, so the search can concentrate on a specific region of the Pareto real front, which is usually the central area of this. However, the interest of any multiobjective genetic algorithm methodology is to find well-distributed border, so the search should be expanded beyond that only the central zone. This is why MOEGA makes use of the crowded operator proposed by Deb (2002) for its NSGA-II algorithm and mentioned in section 4.2.6 both to find the evaluation function of each individual and to select the individuals who will be part of the next generation.

Performance Measurement – Aptitude

MOEGA gives priority to individuals that more quantity of individuals dominate the same those who are more distant from the rest of the population, in order to find a border better distributed and closer to the real.

To fulfill this purpose, it is proposed that the performance measure for each determined by the following equation:

$$f_i = Q_i*(1+D_i)^2$$

where Q_i determines the number of individuals that dominates i and D_i represents the factor of distance from this individual.

In the first iteration, both the number of individuals that dominate and the operator crowded are from individuals already repaired, and from the second iteration, where already exists an external file called Elite, they are taking in counts both the individuals of the elite and the present generation.

External file ELITE

In the first iteration is created an external file called Elite of constant size, where the best solutions found in this iteration are stored. From the second iteration this file is updated with the best solutions that are find. This is a concept that is taken from SPEA2 (Zitzler et al., 2001), and the form how the file is filled is a proposal of this research.

Initially the elite will not accept repeated solutions. The first individuals to do part of it are the non-dominated individuals, then if the size of the elite is not completed, the file is filled with the less dominated non-repeated individuals, and if finally still the file is not completed, then it is filled with less individuals dominated.

It is proposed that individuals who leave the elite to be dominated by some individual, become part of the present generation by replacing the individuals who went out to shape the current elite individuals belonging to the elite at the end of the MOEGA algorithm run will be which are finally shown to the decision maker as possible solutions.

Parent Selection

Parents are selected from individuals belonging to the elite and the current generation, imitating the concept proposed by SPEA (Zitzler and Thiele, 1999) and selection is made by probabilistic tournament, described in section 3.3.2.2, where it is proposed which competes a percentage of the total of possible parents.

Crossing and Mutation Operators

Given that in real-life problems we work with a large number of variables of decision, then it is proposed to make use of the cross in two points, since as was said in section 3.4.1 crossing at a point brings disadvantages in this type of problems.

The mutation operator varies slightly compared to that described in section 3.5, since it is proposed not to find a probability of mutation, but rather a number of variables to be mutated, which will be randomly selected. This with the end of minimizing computational time.

In addition, taking what was proposed by some researchers and mentioned in (Coello 2004), it is decided to mutate a greater number of variables at the beginning of the process and to decrease this number as the number of generations advances.

Selecting the New Generation

Taking the concept of NSGA-II (Deb et al., 2002) of bringing parents and children together the new generation, it is proposed that MOEGA group the current generation with the new offspring and from them selects the new population as described by (Deb 2002).

For it organizes both the current generation and the children in layers of non-dominance, and creates the new generation with individuals from the first undominated layers; at case of the last layer accepted, choose the most distant individuals with the help of crowded operator, as described previously.

MULTI-OBJECTIVE MULTI-TARGET GENETIC ALGORITHM INCORPORATING PREFERENCES OF THE DECISION - MOEGA-P

Incorporating human preferences has been a very important topic in optimization evolutionary multiobjective, since the ultimate goal of an optimization problem is not to find the whole Pareto frontier, but to find an alternative as a solution that is which the decision maker would eventually use. However, it is not useful to find a single solution, since the decision maker might feel somewhat uncomfortable and insecure with this one, since know that there are many more equally efficient, and could not have the option of know them unless you rerun the optimization algorithm.

With the interactive methodology presented in this section, it is possible to comply with these two conditions, since the decision maker when adjusting his preferences during the algorithm finally finds a small fraction of the border of Pareto, in which very surely will be your preferred efficient solution. In addition, this methodology takes into account certain characteristics that must be consider when you want to work with preferences, such as (Coello, 2000):

- The perception of a person will always be influenced by the elements that are at a given time and by the environment in which this moment occurs.

- A person's preference function or value structure can not be however, it can be assumed that the decision maker has a set of beliefs attached to it.
- The preferences of the decision maker are changing over time depending on their structure of value.
- After a person acquires experience and with it an apprenticeship, their aspirations and desires for achievement vary.

To achieve this, MOEGA-P allows the decision maker to prior to defining their preferences, can perceive what is really happening with the solution of the problem and that later learning, he is free to define the consider it necessary. Before proceeding to explain in detail the proposed methodology, determine how the preferences of the decision maker will be represented. In the literature it was observed that these have been included in the following way (Coello, 2000), (Rachmawati and Srinivasan, 2006):

- By weights given by the decision-maker representing the interest of a target with respect to the others. These weights are used in a scalar function with which parents are selected.
- By the minimum level required for each objective. Individuals who do not comply with the minimum level in a specific objective, are penalized in their function of evaluation.
- By a reference solution, that is, a numerical goal that you want to meet the decision maker. The parents selected will be those closest to the point of reference or goal.
- Finding the relative importance of objectives through linguistic variables and then converting these into weights or intervals of weights that are then used within an aggregating function for parent selection. In the case of intervals, choose a weight contained in it.
- Finding a utility function that expresses how the decision-maker combined the objectives, order to prefer one solution to another.
- Choosing between two alternatives, based on compensation rates between objectives. These rates are based on the points obtained in the algorithm.

- Classifying a set of solutions instead of classifying the attributes.
- Defining previously qualitative goals with the help of fuzzy logic.
- Defining two sets of weights: The maximum and minimum that would be willing to sacrifice of a goal. With these weights we add linearly the objectives Weighted and we work with Pareto dominance.
- Combining multiple goals with priority between objectives, making use of the AND and OR (Diffuse Logic) operators.
- By means of an indicator that measures the quality of the solution, for example robustness.

Considering that the decision maker can express preferences through weights, it could be ideal situation, since it is difficult for a person to express preference in an explicit number of absolute form, in addition, researchers have demonstrated (Coello, 2007) that the use of weights through aggregative functions to solving multiobjective problems does not work satisfactorily.

On the other hand, define utility functions in a progressive way to represent the interests can be cumbersome, as well as selecting or classifying solution in the course of the algorithm, since it would require a great effort mental. However, the use of fuzzy logic to represent preferences is an area quite interesting, since it assumes that the decision maker does not know exactly what his preferences, but has a global idea of them. However, in this this resource will be used, but it is considered as future work.

Because it is sought that the decision maker does not have to make a greater effort at the time of your preferences, since you must do so on several occasions during the algorithm, MOEGA-P proposes that it represent them by means of a value that provide security at the time of defining it. This value could be your expected minimum. The author considers that a person will always be more secure than it is definitely not willing to accept, more than what we would like to achieve, because generally the latter tends to infinity if it is speaking of gains or to zero if it is talking about costs.

For example, if a person had to decide on choosing a project between a set of them, taking into account the profits they generate, it would be easier for decide when you really know that some of these projects do not meet the minimum profit required by him, so he would never be willing to execute them. That is to say, it will always be easier for the decision maker to decide what he does not want to lose, rather than what you would like to win. However, these expected minimum values vary as the decision maker learns from problem, since initially the decision maker may be more discreet when it comes to define the minimum you want to get, but when you are presented with possibilities each more attractive according to his interests, he is increasingly demanding in his values minimum, or may even become more cautious.

Therefore, these minimum values must be dynamic within the search process. Before making a decision on minimum values, the algorithm runs several iterations with MOEGA without taking into account the preferences, this in order that the decision maker has a first approach to the problem and feels more secure at the time of set your preferences for the first time; the number of previous runs will be a percentage of the total runs of the algorithm. After setting preferences, the search will be concentrated in more restricted areas, it is therefore considered necessary that the can make decisions about the exploration and / or exploitation of the search, for it`s decision maker can modify the parameters of probability of crossing and the percentage of variables to mutate as the algorithm progresses.

After making decisions about the minimum values allowed and the exploitation and search exploration, the algorithm works with these new values and the decision maker can visualize through graphs the behavior that have had the alternatives of the elite and those of the total population; At the same time you can graphically see the range of the possible values that each target takes in the whole population, in this way the user would be careful not to require the problem to give a solution that may be out of reach. In addition, you can view an average of exchange between its objectives, in order to assess whether or not it is in the interest of minimum of one of its objectives. With this information, the decision maker may have an idea of how much requirements with respect to an objective if it wished to increase the minimum expected value of another one of them; or if it is not necessary to diminish its exigencies but on the contrary still is increase each of them.

The decision maker may pause the algorithm at any time redefine your preferences.

The expected minima would be expected to limit the search area by making it decreasing as the process progresses and therefore the search becomes much easier and interesting for the decision maker since it would be guided Pareto fraction of their interest, in addition to this, would improve the efficiency of the process from the point of view of computational cost.

This perception of preference will be introduced to the algorithm proposed in section 5.2 incorporating a new key concept called "Factor of Preferences", which modifies the dominance and with it the function of evaluation and the selection of a new generation. It also modifies the way individuals are accepted into the elite.

Preference Factor

In order to evaluate each solution alternative with respect to the preferences of the decision maker, it has been found necessary to find a preference factor degree of compliance with the expected minimum values for each alternative i.

Because the minimum values remain somewhat uncertain for the decision maker, ie, that are defined with a 100% certainty, then it is proposed to find the preference by a fuzzy function, as shown below:

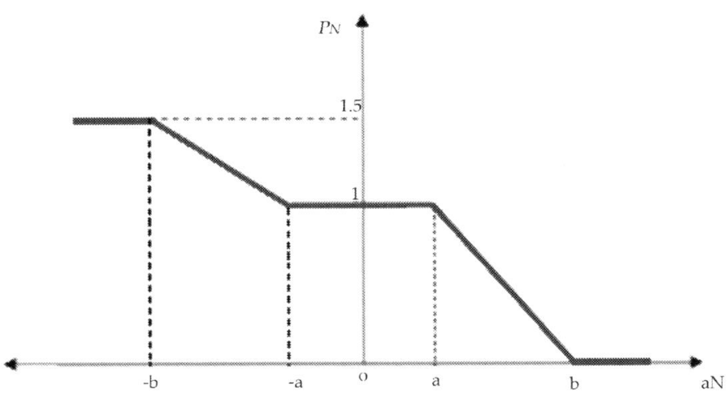

Figure 5.1: Function of Preference Factor.

Thereafter, the following notations are used:

d_i: It is the difference between the minimum allowed by the decision maker and the value alternative i in a given objective. This value is negative when the value of objective of the alternative exceeds the minimum required by the decision maker.

p_{ij}: It is the preference factor of alternative i when the objective j is evaluated.

a: It is a value of indifference for the decision maker where the objective of the alternative that is evaluating can be at that distance from the minimum required by the decision maker and be assuming that you meet your preferences 100%.

b: It is the maximum difference allowed by the decision maker to say that an alternative fulfills in some way their preferences, that is, when the difference exceeds this value the decision maker can say with 100% certainty that it does not meet their preferences and therefore is not interested.

With the rank between a and b the decision maker can express his degree of exigency for the your preferences, being more demanding with a small range and more flexible to the extent that this range is broader.

The preference factor should represent that it will be much more important than a individual to fulfill his / her preferences in all the objectives, that he / she fulfills them in great way in a few and in others violates them, that is why the function will punish a lot more than it will award, even if the amount that exceeds the minimum expected value is the same one in which it violates. Therefore, ij p is in the interval [0.0, 1.5] where it takes values between [0.0, 1) when the alternative fails to meet the preferences given by the [1.0, 1.5] when it is equal to or exceeds the minimum required value.

The values of a and b, as well as the expected minimums, are given by the decision maker after having perceived the scenario in which it is and can be modified during the run of the algorithm, so it could decide whether to be more demanding or not in the first or last races.

Finding the values of ij p for each objective j, then the preference as follows:

$$FP_i = \frac{\sum_{j=1}^{m} p_{ij}}{m}$$

where m is the number of objectives.

Dominance by Preference

The concept of dominance for an individual is probabilistic, where the alternative dominates with a probability to the alternative j, if alternative i is equal in all the alternative to j and is greater than alternative j in at least one of the objectives, with probability being the preference factor of the alternative (FP_i).

Thus, an individual i with FP_i: ≥ 1 and who dominates another by its target values, then an individual with FP_i: <1 who dominates another by his goal values, you will have a probability of failing to be effectively dominant, likewise an individual with FP_i=0 will never be dominant despite its values in the objectives.

In the case of Figure 5.2, where we have two objectives, and assuming that FP_1= 1.2, FP_2 = 0.85 , and FP_3 = 0.4 , alternative 1 will always dominate alternatives 2 and 3, and alternative 2 has a probability of 0.85 to dominate alternative 3.

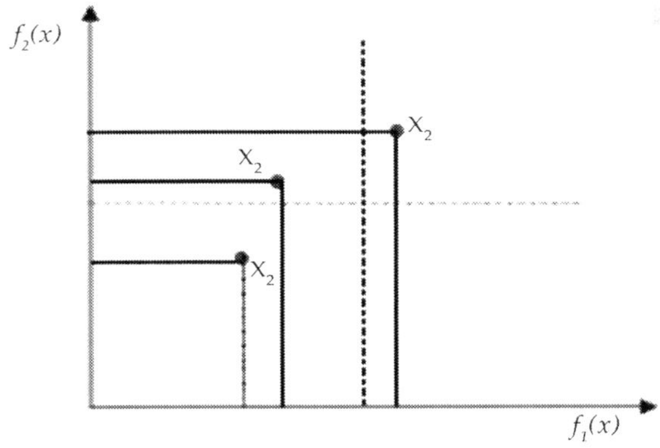

Figure 5.2: The dominance of alternative 2.

The dominance of alternative 2 is not deterministically accepted since the latter does not meet the decision maker's minimum expected value for objective 1 and letting you be dominant to this alternative is to allow it to continue reproducing in the next generations and bring the population to areas that are not of interest to the decision maker.

This would give more importance to those individuals who actually fulfill or are very close to meeting the preferences of the decision maker.

Evaluation Function – Aptitude

In section 5.2.4 the evaluation function for MOEGA depends on the distance factor of each individual and the number of individuals that each one dominates.

In the case of incorporating the preferences of the decision maker, it is not necessary for individuals are as dispersed as possible because of the search therefore the distance factor will not be used. However, the number of individuals continue to be part of this calculation, taking into account that now the concept of dominance has changed.

It is given that preferences within dominance are assumed to be only one probability, at the moment of defining the attitude value for each individual is necessary to take them into account, in this case the preference factor does the times of punishment or prize to the evaluation function depending on whether or not to violate the preferences of the decision maker.

An individual among the most violates preferences will have a lower probability of being reproduced than one that less violates them. And if not, an individual among meets the preferences will have a greater probability of being selected as a parent.

Now, the evaluation function is given by the following equation:

$$f_i = Q_i * FP_i^2$$

Select the New Generation

It should be remembered that in the Elite will be kept the individuals who will finally do part of the solution presented to the decision maker, so it is important that in this file remain those interesting individuals for him.

As this file has a fixed size, it is now completed by the following individuals and in the same order:

- Individuals not dominated, not repeated and also fully complying with the expected minimum values.
- Individuals less dominated, not repeated and also fully compliant with the expected minimum values.
- Individuals less dominated, not repeated, no matter that they do not fully comply with the expected minimum values.
- Less dominated individuals, regardless of whether they are repeated or not meeting totally with the expected minimum values.

In the case that the first group of individuals is much larger than the maximum size of the elite, then the individuals with the highest value should be selected from them. In the last iteration, only those that of the elite fully comply with the preferences of the decision maker.

The new generation is selected by non-dominance layer, as proposed by MOEGA, considering that the concept of dominance varies.

Because it is difficult for the decision maker to define his preferences deterministically, MOEGA-P gives you the possibility to re-evaluate them whenever you deem it necessary, not without before allowing you to visualize the panorama in which it is.

MOEGA-P, uses as basis the proposed algorithm MOEGA, however due to the management that are given to preferences, this same methodology could be used in others algorithms that work with the concept of dominance, such as SPEA2 and NSGAII.

Since it is necessary to evaluate the proposed methodologies, the following choose a problem combinatorial test, which will be applied to the two algorithms and to those proposed in this research in order to evaluate their contribution to the state of the art.

RESULTS OF THE SIMULATION

In order to validate the methodologies proposed in Chapter 5, a problem Combinatorial test well known in the literature called Problem Multi-objective of the Backpack and this will be solved with each methodology for 2 and 3 objectives varying the number of possible articles between 400, 600 and 800 articles.

To determine whether or not MOEGA obtains closer Pareto frontier results with a uniform distribution along the boundary found, the Generation Distance - GD and Spacing-S metrics (Coello et al., 2002), and compared to the values they take for the NSGAII and SPEA2 algorithms, from this if the proposed algorithm can obtain better solutions than the reference algorithms.

In the case of MOEGA-P, it will be validated solving the same problem but for two objectives, in order to graphically show the results on a Cartesian plane and to show if it is really necessary for the decision maker to incorporate his preferences during the algorithm and not before or after it.

The problem combinatorial test and the metrics that will be used to measure the efficiency of the new methodology – MOEGA. Then the configuration used for each of the algorithms and finally the results will be shown for both MOEGA and MOEGA-P, defining the utility of each of them.

EVALUATING PERFORMANCE OF MOEGA AND MOEGA-P

In order to evaluate the proposed methodology, a problem of classic optimization called Multiobjective Problem of the backpack (Multiobjective 0/1 Knapsack Problem) which is well known in the literature and qualified by (Zitzler and Thiele1999) as a general problem, understandable, easy to formulate and represents a type of problem in the real world.

This problem basically consists in selecting a number of articles that have associated with them a gain and a size. These articles will be backpack which has a limited capacity and seeks to maximize the total gain obtained when choosing the items. Initially this problem was designed for a single objective and then was adjusted for multiple objectives referring to several backpacks.

The formulation of this multiobjective problem is defined as follows (Zitzler and Thiele, 1999):

Let n be the number of articles and m the number of rucksacks.

c_{ij} = Gain of the item i put in the backpack j

a_{ij} = Size of the book i put in the backpack j

b_{ij} = Capacity of the backpack j

The solution vector $x=(x_1, x_2, x_3, \ldots x_n)\{0,1\}^n$ which takes values of 1 when the book is selected and 0 when it is not, such that:

$\mathrm{Max} f(x)=(f_1(x), f_2(x),\ldots,f_m(x))$

$$f_i(x) = \sum_{i=1}^{m} c_{ij} x_i$$

Subject to:

$$\sum_{i=1}^{n} a_{ij} x_i \le b_j \qquad \forall j \in \{1,2,\ldots,m\}$$

This problem will be solved for two and three backpacks, in order to evaluate the methodology for two and three objectives, varying in each of them the quantity of articles between 400, 600 and 800. The data used were calculated in the same way as proposes in Zitzler and Thiele (1999) with the difference that it will be demanded a little more, since increase the range of values for the earnings and the size of each item and the total capacity for each backpack will decrease. The profits and the size of each are given by random numbers between 10 and 150 and the total capacity for each backpack will be 40% of the sum of the sizes of the items in each backpack.

METHODS OF EVALUATION

Two key parameters that need to be taken into account when evaluating a Multiobjective genetic algorithm are the closeness of their results to the Pareto frontier real and the distribution of these through it. At present there are many metrics that allow to compare several algorithms in relation to these two criteria, which are mentioned in Coello et al. (2002). In this research, two of them will be used.

Generational Distance (GD)

GD represents on average how far a non-dominated border of the border of Pareto. This metric is obtained from the following equation:

$$GD = \frac{\left(\sum_{i=1}^{n} d_i^p \right)^{1/p}}{n}$$

Where n is the number of individuals found at the border, p = 2, and i d is the euclidean distance in the space of the objectives between an individual of the border found and the closest of the actual border.

Note that when d_i is zero, it means that the individual found matches that of the border, then, as long as this index has a value closer to zero, it will be better, since it would indicate that the border is closer to the real one.

Distribution (Spacing - S)

This metric represents how uniformly distributed individuals are to length of the border found. Is defined as:

$$S = \sqrt{\frac{1}{n-1} \sum_{f-1}^{n} \left(\bar{d} - d_1 \right)^2}$$

Equivalent to the sample standard deviation of the i d. Where n is the number of vectors at the non-dominated border, i d is the distance from the individual i to the individual j more which is defined for two purposes as:

$$d_i = \min_j \left(\left| f_1^i(\vec{x}) - f_1^j(\vec{x}) \right| + \left| f_2^i(\vec{x}) - f_2^i(\vec{x}) \right| \right) \qquad i, j = 1, \ldots, n$$

And

$\bar{d}$ is the mean of all d_i. In the case of a problem with m objectives, the author will redefine the value of i d as follow:

$$d_i = \min_j \left(\left| f_1^i(\vec{x}) - f_1^j(\vec{x}) \right| + \left| f_2^i(\vec{x}) - f_2^i(\vec{x}) \right| + \ldots + \left| f_m^i(\vec{x}) - f_m^j(\vec{x}) \right| \right)$$

When this metric takes the value of zero, it means that all individuals in the border are distributed equidistantly.

These two metrics will be used to evaluate if the proposed algorithm MOEGA improves in the two criteria mentioned above with respect to the NSGAII algorithms and SPEA2.

ALGORITHM CONFIGURATION

In MOEGA a varied mutation is proposed, since after studying the graphically the behavior of the population during several races it was noticed that better results when greater exploration is allowed at the beginning of the iterations and decrease this at the end. The author argues that this is because, as the solution is close to the Pareto real border, the restrictions will be less, so if you have more exploration will be more difficult find good solutions within the feasible area, but greater exploitation will that solutions are found close to those already found and much better.

The number of iterations before defining preferences in MOEGA-P, will be 10% of the total of iterations. He works with a constant Elite of 80 individuals throughout the and only at the end of it, the elite will only receive individuals who are not dominated and which also fully meet the final preferences of the decision maker.

In order to compare the algorithms, it was necessary to program NSGAII and SPEA2. This is done as described in Deb et al. (2002) for NSGAII and in Zitzler et al.(2001) for SPEA2. Both MOEGA and MOEGA-P were programmed in C # with Microsoft Visual Studio 2005.

CONCLUSIONS

The proposed multiobjective genetic algorithm (MOEGA) is much closer to the border for the problem of the backpack with 2 and 3 objectives, compared to the methodologies NSGAII and SPEA2. However, the distribution of their individuals is not so uniform as that obtained by SPEA2, but much more than that achieved by NSGAII even though MOEGA uses the same NSGAII procedure to distribute its individuals.

In order to improve this distribution, one could use the methodology proposed by SPEA2 and analyze the results.

When studying the behavior during several runs of NSGAII, SPEA2 and MOEGA in the problem of the backpack, it was observed that these multiobjective genetic algorithms tend to focus on the central area of the Pareto border, could offer the decision-maker solution alternatives located at the ends of the border, which may be of interest to him. Directing the search interactively could solve this problem, as results were obtained that were not possible obtain them without intervening in the process.

Another advantage of MOEGA-P is that it limits the search in areas of interest to the decision maker, it was possible to use the same computational cost in smaller regions to reach levels closer to the real frontier, so it can be concluded that it is desirable to involve the decision maker's preferences during the search and the manner in which MOEGA-P does.

After the development of this investigation the following can be concluded:

- The scientific community of multiobjective genetic algorithms has focused its greater attention in the algorithms that find all the border of Pareto and better distributed, without completely solving the decision problem, since with the solution found with these algorithms would continue with a decision problem discreet.
- The methodology proposed in this investigation (MOEGA) achieves better results than those obtained by NSGAII and SPEA2 when solving the problem of backpack for 2 goals. In the case of three objectives it is not possible to distinguish which of the three methodologies achieve a closer approximation to the real border, but MOEGA is within reach of the other two methodologies that are recognized within the state of art.
- With regard to distribution, MOEGA when it is solving the problem of backpack with 2 lenses, you get most of the time, individuals with a more uniform distribution than those obtained with the other two methodologies. Do not the same happens for the case of three objectives, since SPEA2 has better distribution, however, if it is much more uniform than that achieved by NSGAII yet when MOEGA uses the same procedure to distribute its individuals.

- Even the so-called state-of-the-art algorithms (NSGA-II, SPEA2),applied to the problem of the backpack, can throw solutions that do not understand those of interest to the decision maker, since it was found that these algorithms tend to find all its individuals very close to each other concentrating their search in an area of the Pareto border, which in this case was the central one, forgetting the possible solutions located to extremes and that could be of interest to the decision maker.
- The use of preferences in these methodologies for the backpack problem, allows the decision-maker to contemplate some of these points as possible solution, since you can direct the search to that zone. The preferences of the decision maker are changing over time depending on the scenario in the one that is, as well as its aspirations and desires of achievement vary in the measure learned from the problem, that is why it is risky to consider preferences are absolute and that once defined he would not be interested to change them again. Therefore, it is necessary to give the decision-maker the option to modify preferences as you learn from the problem and at the MOEGA-P.
- MOEGA-P with its way of including the preferences, for the problem of the backpack, that the decision-maker can obtain alternatives that are not possible with a without preferences, allowing you to actually to their interests.
- By limiting the search zone with the preferences of the decision maker, it wasting computer costs by looking inside a large area when perhaps the solutions found in it will be discarded because it is not in the interest of the decision maker. Therefore, this computational cost MOEGA-P uses it to find solutions more efficient in the desired portion of the border.
- As the algorithm with preferences advances and finds solutions near the border of Pareto, the search space will be increasingly constrained, which leads to the decision maker needs to modify his or her exploration and exploitation so that their search becomes increasingly efficient, since that if it continues with those initially established, the algorithm would lose time in seek solutions outside the restricted area.

Some of the works that could be developed after this are:

- Propose a new concept of distribution of individuals for algorithms multiobjective genetic algorithm that is effective from the beginning of the search, the population is dispersed throughout the area bounded by the restrictions.
- Propose self-adjusting crossover and mutation rates depending on how dispersed is the population in determined iteration.
- Construct new metrics that determine whether an algorithm approaches the border of Pareto and that is distributed along it.
- To propose a methodology where the decision-maker can define the sacrifices among the objectives through linguistic variables and that the concept of dominance is considered in a diffuse way.
- Make use of artificial intelligence techniques such as neural networks or inference to train an agent to make preferences during the search process of the multiobjective genetic algorithm.
- Construct methodologies that take into account multiple decision-makers at the define preferences.
- Apply MOEGA and MOEGA-P to real problems, evaluating the satisfaction of the decision-maker with respect to the solutions found.
- Improve the application, so that the information can be provided to the decision maker when solving a problem with more than two objectives.

CHAPTER 6

ADVANCES IN GAME THEORY

Ivan Stanimirović

INTRODUCTION AND HISTORICAL DEVELOPMENT

German matematician Ernst Zermelo in 1921 proved an important assertion of the existence of winning strategies. John von Neumann was defined back in 1928 and employed some concepts from game theory. However, it is the emergence of new theories remain outside the scope of professional and very little has been done in this area to the appearance of the book. In this book the first time the task appears with numbers. The foundations of game theory laid the John von Neumann and Oskar Morgenstern in 1944 in his book "Theory of Games and Economic Behavior".

Then the final date mathematical basis of the theory, defines the basic concepts of strategy and value games, formulated a method for finding optimal strategies and pointed to the possibility of applying this theory in economic research. Dz. Dantzig in 1951 showed the connection between linear programming and game theory. More specifically, it has proven to be a special case of the game, "game with zero sum" can be reduced and solved as a linear programming problem. Since then began to grow interest in the development and application of game theory. It has become obvious that the logical structure of the theory of games hides a lot of options in terms of decision-making in the economy. Another fundamental contribution to the work of John Nash, for which he was later awarded the Nobel Prize for economics (Nobel Prize is not awarded for mathematics). About John Nash was recently recorded biographical film A Beautiful Mind. In the beginning, the main motivation for the development of the economic, or the game theory found application in other social sciences, military applications, and even applications in biology. Interesting examples from biology can be found in the book by Richard Dawkins, The Selfish Gene. A special aspect of the competitive situation of participation in the auction (bidding) for contracts, concessions, rights and different. The essential conflict in a situation offering the following: how many deals increases the chance of winning the job (lower price bids) until the same time reduces the amount of profit and vice versa. The problem then is how to achieve the best balance between chances of getting an offer on the one hand, and achieving satisfactory profits on the other. In the context of game theory models have been developed that provide assistance in solving this type of problem.

BASIC CONCEPTS AND PRINCIPLES OF GAME THEORY

Basic concepts that characterize the theory of games are:

- Players: are the parties to the conflict;
- Game: model of the real behavior of participants in a conflict situation;
- Gain (loss) represents the result of the game;
- zero-sum game: the type of game in which one party receives as much as the other side loses;
- set of strategies (move, alternative) represents the behavior of each of the players (that's the plan development of the game);
- optimal strategy: a strategy that numerous repetitions in the game provides the player the maximum possible gain medium, medium or minimum possible loss.

The criteria, which are used to select the optimal strategy can be numerous:

- *min max/min max* criterion (Wald, pessimistic);
- *max max/min min* criterion (optimistic);
- Hurwitz's criterion (method of optimism-pessimism);
- Laplace's criterion (principle of insufficient reason);
- Savage's criterion; (Methods minimax regret);
- The criterion of expected value;
- Bernoulli's criterion; etc.

The top five of the listed criteria are applied to the conditions of uncertainty, while the expected value criterion applied in terms of risk.

Min-max criterion is the main criterion used in game theory. Also called the Wald criterion by A. Wald who found a rule which is based on this criterion: "If my unknown state of nature, the occupation of the most conservative stance." This criterion supports the cautious and pessimistic attitude, but without going into the risk tends to limit possible loss.

Optimal strajegija under this criterion is this strategy which, in multiple repetition game provides the player the maximum possible gain medium, medium or minimum possible loss.

Starting from such rules can be defined the following principles:

- The player chooses their behavior so that his winnings in the game is the maximum with the least favorable default behavior of the opponents. This procedure is actually a selection of the most cautious strategy, and the principle is called maxmin principle.
- The player can choose their behavior so that his losses are minimal when, according to him, the most unfavorable functioning of the opponent, where this principle is called MinMax principle.

These principles suggest a cautious approach, which differs from the dice game theory.

The term "game" is used as a scientific metaphor for a wide range of human interactions in which the outcome depends on the mutual interaction strategy two or more parties who have conflicting interests. The main issues raised in game theory are:

- What does it mean to choose "rational" strategy, where the outcome of the strategy depends on the strategy chosen by the opponent and when the information is incomplete?
- Is it rational in games that allow the mutual gains (or mutual loss) on both sides, mutually cooperate to ensure the greatest mutual gain (or hit the smallest mutual loss) or rational act aggressively, regardless of the possibility of mutual gain or loss?
- If it is sometimes rational to play aggressively, which are situations in which it's worth it, and in what situations it is rational to cooperate?
- Whether from the process of interaction arise spontaneously moral rules of cooperation rational egoist?
- What is the real attitude of people to the "rational" recipes offered by game theory?
- If the real behavior differs, in what direction is this? Are people more cooperative than it proposes to game theory? Or are more aggressive? Or both?

Game theory can be used to analyze hazardous games such as poker, or in the analysis of sports games, such as football, but also some phenomena that are not games in the previous terms, such as market competition, arms race, environmental problems and some quite exotic situation, as a matter of knowing that if there are supernatural beings and many others. Each player has a choice of appropriate strategy, the outcome of which depends on the strategy chosen by the other participants in the game. Thus, the problem of game theory is a mathematical problem of maximizing profit group or single player, which is a rational solution of the game.

For practical use of game theory, it is interesting to examine the links with economic theory. The key link between game theory and neoclassical economics, the concept of rational behavior. Neoclassical economics is based on the assumption that human beings are absolutely rational in relation to the possibility of economic choice. In other words, each individual tends to a given situation maximizing your winnings - profit or other subjective benefit.

The concept of a rational individual who chooses their strategies so as to maximize the expected benefits is definitely an approximation of the real behavior of human beings, which can sometimes be irrational, and depends on the physiological characteristics of each individual person. However, this approximation is proving to be good enough to explain, on average, people's reactions when confronted with specific economic choice.

The assumption of rationality has a dual role in the allocation of research resources. First, it narrows the set of possible solutions, because absolutely rational behavior is more susceptible to analysis than irrational and, secondly, it provides criteria for assessing the efficiency of the economic system. If the economic system allows each lose more than others get (costs exceed benefits), then something is wrong with such a system, as is the case with environmental pollution, over-exploitation of natural resources, lack of investment in scientific research and the like.

In neoclassical economics, rational individual works within a specific system of institutions, such as private property, money and free markets. The consequence of this is that the individual does not have to analyze your interaction with other individuals, but only his own position and "market conditions". Whenever competition is limited (but there is a monopoly), or when private property is not well regulated by the law, can not be applied neoclassical economics. Game theory is not intended to resolve such situations, to provide a theory of economic behavior when individuals interact directly rather than through a "market mechanism".

UNCERTAINTY GAMES

Uncertainty results of the game stems from three basic causes:

- Rules of the game are such that there is a very large number of variants of its unfolding, so it is impossible to accurately predict the outcome of games. This kind of uncertainty is called combinatorial, a game in which it is the main cause of the uncertainty of the result is called combinatorial games. A typical example of this type of game is a game of chess. With the advent of computers, this type of games can be solved so as to define a set of winning combinations and in principle obtain a satisfactory solution.
- Another source of uncertainty is the presence of random

factors. Games in which the dominant cause of the condition the uncertainties referred to as hazardous, such as various dice games, roulette and the like.

- The third source of uncertainty is the absence of information about possible actions that opponents of his strategy. Games of this type are called strategic.

THE DIVISION GAMES

The game can be defined as a set of rules governing the flow of the game and the player's behavior. Players participating in the game have to know the rules of the game and must observe them.

Games can be shared by different criteria.

A. One of the criteria for the division of mathematical games based on the presence or absence of elements of randomness. According to this principle, the game is divided into two different types: games of chance in which the player your abilities can influence their outcome, and strategy games where the outcome of the game affect the ability of individual players. The elements of chance are particularly pronounced in hazardous games such as card games or dice. For formal study are particularly significant strategic games.

B. According to the number of players the game can be divided into games with one player, two-player games, and games with an arbitrary number of players. The term "player" should be understood in very general terms, as well as the very concept of the game. This is the number of opposing sides in a game (individuals, groups, competing participants, organizations, armies, etc.). Here we deal with games with two players. The famous game of chance in which one person was involved plays solitaire. In this game are equally important how the skills of players and players luck factor. The most famous game with two players did chess. Games involving many players as poker and bridge.

C. Each player in the decision-making has a number of options, or alternatives, among which selects one he thinks is the best.

According to the number of alternatives that a player has available, the game is divided into finite and infinite. The game is final if it has a finite number of positions and if each party and the game ends in a finite number of moves. We will study the games in which each player has at its disposal a finite number of alternatives. In connection with this split games you need to define a strategy, as one of the basic concepts in game theory. Each of the players at the beginning of the game, and assesses their opponent alternative, weighed all the possible combinations and selects the one that seems best to him. In this way, each player chooses an alternative that at any point in the game is pre-determined. The strategy is defined as a sequence of decisions that the player is able to play to make.

D. The games can be divided and according to whether the players have complete information on the position of the game as well as the possible moves at each position (about those of moves of your opponent). According to this criterion games are divided into games with complete information on games with incomplete information. Chess is a game with complete information. Bridge and other card games are examples of games with incomplete information. Here are studied the game in which both players have complete information.

E. The games can be divided according to the characteristics and the price of the game, and the result of the game. Each game ends with a certain outcome, which is expressed profits or some estimates. After each Game - election by a strategy by each player, performs the calculation and payment. A player who loses pays the player who receives the amount of which is determined by the rules of the game. There are games where at each such calculation sum of gains of one player is equal to the sum of paying another player. These games are called games with a score of zero (zero-sum game) or antagonistic games. In addition to these, there are also games with nonzero sum, but we will deal with the game with a score of zero.

Based on the division games, it can be concluded that the field of operations research interests primarily the study of strategy games with a finite number of alternatives involving two players who have complete information, and that there is no element of chance. Such games satisfy the following conditions:

- The game involved two players (first and second). In the game there is a finite number of positions, where one of them is chosen as the initial.
- The game has clearly defined rules which allowed certain (legal) moves players in each position.
- The first and second player withdrawing their moves in turn.
- Both players have complete information.
- The game was equal, which means that in every position possible moves does not depend on who the shooter. Chess is an example of a game that is not equal, because in every position first player plays only the white pieces and the other player plays only the black pieces.
- The game is final, that is. each party and the play ends after a finite number of moves, when one of the players, which is on the move, he does not have legal move. In a normal game, the player who finds himself in such a situation - losing. In the reverse play, a player who finds himself in such a situation - obtained. The game can not be unresolved due to repetition of moves.
- The games end result is zero.

F. Depending on the relationship between players of the participants, all the games are divided into cooperative and non-cooperative. Cooperative games are that kind of games in which players form coalitions that serve them for mating behavior and choice of individual strategies, which give them achieve the most favorable results. If during the game there is no coordination in behavior (choice of moves) by the player, such noncooperative game represents, ie. antagonistic game.

LOST AND WON POSITIONS

For the players (game participants) strategy is a set of instructions for playing that contains instructions what to do in every situation that may occur during a game. Mathematical models of game theory more difficult finding the optimal strategy player guarantees the highest possible average win per game.

The term "mathematical game" is contradictory in the following sense: at a time when a game is solved mathematically, ie. When the game gets calculated by the algorithm, it ceases to be a game. However, as you know from the history of many games fully trained, but who have they lost their vitality. They play as a challenge to those who are ignorant of the game to reveal the "secret" of victory.

Although most games consist of a series of moves that alternately pull the two players, sometimes a series of moves each player can define some rules. Such rules define the behavior of the players in each of the possible positions, ie. defines the strategy of the game. For more complex games such as chess or bridge, for now such a reduction only in the domain of theoretical, but not practical options.

In many games the player can be in certain positions to come to the conclusion that it can win regardless of the moves of his opponents. Then it is said that the player has a winning strategy. An important goal of game theory is to characterize obtained and lost positions, as well as to define the methods for finding a winning strategy in the final games.

Consider the following examples from which to draw certain conclusions.

Example 6.5.1. On the table are two piles with 7 chips. In one stroke taken an arbitrary number of chips (at least one), but only with a crowd. The winner is the player who picks up the last token. Find a winning strategy for the second player.

Solution. Another player provides a win using a strategy symmetrical: each turn he takes as many tokens as much as in the previous move took the first player, or the opposite of the crowd. Symmetry is here reflected in the number of chips equal on both sides of the mound. In this way, the first player must with every stroke impairs symmetry. As the finishing position symmetrical, it can achieve only other player.

Example 6.5.2. Players alternately placed one jumper at a chess table fields so that they are not mutually "beat". Get a player who cannot move to withdraw. Which player has a winning strategy?

Solution. Winning strategy has the second player. It is enough that each turn set jumpers on the field that is symmetric with respect to the field in which the first player to put the last jumper. On seldećoj picture shows the panel dimensions 8×8. On it are two symmetrical fields on the board the same color, so that the jumpers on these fields is not «beat». It is clear that the other player has a move available until the first player has a move. In Figure 6.1. is shown a possible jumper arrangement, whereby the number 1 marked the first player moves, while the number 2 marked the second player moves.

Formally speaking, the victorious strategy in these games is coming in the way that all positions are games divided into two categories. Positions one class called winning, while the position of another class called losing. In doing so, the following applies:

- The final position is a loser (in the normal game losing player who is on the move).
- There is a move that from a losing position leads to another a losing position.
- For each winning positions is one stroke can be reached in a loser position.

The move is good if it is translated from wining game in a losing position; otherwise it is bad.

1	2		2		1	1	1
2	1	1		1		2	
	1					1	1
1		2	1	1		2	
	1		2	2	1		2
2	2					2	
	1		2		2	2	1
2	2	2		1		1	2

Figure 6.1. One possible arrangement of jumpers.

In the case of 6.5.1. losing positions are those positions in which the two mounds of the same number of chips.

Solution games can be easily found when ordered classes on the losing position. Then the winning strategy is to move every game from winning positions converted into a losing position. With respect to the property (2), a player from a losing position off his game must be translated into a winning position. Thus, if the initial position of losing (as in the case of 6.5.1), then the second player has a winning strategy; Otherwise, a winning strategy has first player.

Example 6.5.3. On the tape of length n + 1 fields are numbered from 0 to n from left to right. In the initial position of the token in the field with the number n. Each move consists of moving the chips 1, 2 or 3 places to the left. Get a player who may not withdraw legal move, that is. when the token in the field of 0. For which values of n winning strategy has the first player, and that second?

Solution. If the token is in the field 0, losing a player who is on the move. Marked fields 0 - character. If the token is on one of the fields from which it can move one stroke on the field 0, the shooter will lead the chip on the field and 0 win. Therefore, label the boxes 1, 2 and 3 with sign+. Position the field 4 loser because every legal move, the shooter must lead to one of the token box 1,2 or 3, which are marked with +. Continuing the process, we conclude that the fields 5, 6 or 7 position obtained for the player to move and box 8 of the lost positions for players on the move, as shown in Figure 6.2. Thus, the second player has a winning strategy if the number n is divisible by 4, but in all other cases, the winning strategy has the first player.

0	1	2	3	4	5	6	7	8	9	10	11	12	...	*n*
-	+	+	+	-	+	+	+	-	+	+	+	-	...	±

Figure 6.2. Wining and losing positions in a game with coins.

Example 6.5.4. The basic variant of the game "X-O" can be called "Three characters to win." For this game we need 3x3 square. The game is played by two players who alternately written sign "x" or "O". A player who enters the symbol "x" called player - X (or the first player) and the other is called players - O (or another player). The winner is the player who first enroll their three characters in one of the rows of a column or on one of the two diagonals.

As in other games, we need to find the optimal strategy.

- The player who plays first has a positional advantage.

- If no one player does not make mistakes, the outcome of the game is unresolved.
- If the first player to occupy the central field (B2), it may be safe to have the tie, and possibly to win if the other player does not use the optimal strategy. The second player has an optimal strategy that guarantees him a tie. Chances are that the ignition ends in a draw.
- If the first player starts the game on the stroke of one of the angular field, has a good chance to win if the other player does not adhere to the optimal strategy. The optimal strategy is another player gives him a guarantee that they will get at least a draw.
- If the first player started the game on the stroke of one of the four corner box, then it is necessary that the other player his first move, take the central position to reach the tie.

One variant of the game "X-O" is called "Sum 15".

On a sheet of paper draw a rectangle, granted on 9 squares in which they enroll digits 1 to 9. The game is played by two players. One of the players is crossed by a second rounded figures in this box. The winner is the player who first gets three digits of the sum of 15 will surprise us the fact that this game is isomorphic with a variant of "three characters to win." All possible triplets of numbers whose sum is 15 are (1,5,9), (1,6,8), (2,4,9), (2,5,8), (2,6,7) (3,4,8), (3,5,7), (4,5,6), which has a total of 8. We make magic square dimensions 3×3, we will see that each of the 8 triples lie in after one of his row, column, or on one of its diagonals.

Thus, reaching the sum of 15 is equivalent to capturing some of the triplets of numbers that are in rows, columns and diagonals magic square, or playing «three characters to win.»

MATHEMATICAL DEFINITION OF GAMES

Definition 6.6.1. Game is defined as an ordered n-tuple $G = (P, v_0, P_0, S)$, whose components are defined as follows:

- P - a finite set of elements whose position of play,
- v_0 P - starting position,

- P_0 P - set of the final position of the game.
- S - mapping S: P → 2^P, where the 2^P power set set P. Mapping S represents the rules of the game.

The meaning of defined terms is as follows:

For each position x P, S(x) is a set of game positions G (subset P) which can be reached in one move from the position x. If y S (x), then writes x→ y, indicating a move allowed G game which is played from the position x translates into position y. A set is a collection P^0 final position, ie. x P^0 ↔ S (x) = Ǿ. In a normal game, the winner of the player whose stroke is translated game in the final position. In the reverse game, the player whose stroke lead into the final position of the game - to lose. The most commonly observed normal game. Also, usually opposite each game can be reduced to the equivalent of a normal game.

The game can be without end position (P^0 = Ǿ). Then the game can not end in a finite number of moves. If the number of positions in the final game, the game can have an infinite number of moves only if the positions are in the game indefinitely repeated. Positions can be repeated and in games with a final position, and then moves in the game indefinitely repeated.

Development of game G is a set of positions. If the array is final, it can be written as:

$v_0, v_1, \ldots, v_k,$

where v^0 starting position a v_k finishing position. The player who led the game in the final position by winning. The game can then be written as a series of moves:

v_0 ® v_1 ® v_2 ® ... ® v_k,

where v_i ® v_{i+1} represents a move, $i = 0, 1, \ldots, k\text{-}1$. If (in a normal game) to an odd number then the winner is the first player, and otherwise - other.

Definition 6.6.2. The game is acyclic if it is not in one party can come to repeating positions.

If the number of positions in the final game, then it must end in a finite number of moves one player win.

Definition 6.6.3. The strategy of the game $G = (P, v_0, P_0, S)$ is a mapping

$s : P \backslash P_0$ ® P,

in which for each $x \in P \backslash P_0$ is valid $s(x)$ Î $S(x)$. One player in the game G applies to the strategy if at any position v_i play moves v_i ® $s(v_i)$.

Definition 6.6.4. s strategy is winning for any player if he using the strategy to achieve victory regardless of the strategy of his opponent.

Definition 6.6.5. Let N and R positions P subsets that meet the requirements

N È $R = P$, N Ç $R = \emptyset$.

In this case, a set of N represents the core of the game $G = (P, v_0, P_0, S)$ if the following applies:

(N1) for any two of the positions u, v Î N, there is no move u ® v i P_0 Ì N;

(N2) For each position u Î R there is a move u ® x,, where x Î N.

Positions that belong to the core are bad for players on the move. Other positions are good. For the game $G = (P, v_0, P_0, S)$ to move u ® v, u Ï N is good. Otherwise, it is bad. A player who is on the move in a bad position are not available a good move.

MATRIX GAMES

Matrix games are a two-player strategy games with opposing interests. The game ends with a certain outcome, which is expressed profits or some estimates. After each move by one of the players is done the calculation and payment based on a matrix of payment. A player who loses pays the player who receives the amount of which is determined by the rules of the game. Matrix games in which each such calculation summation gains one player equal to the sum of payment other players called the game with a score of zero (zero-sum game) or antagonistic games. In addition to these, there are also games with nonzero sum.

Matrix Zero-Sum Game

Matrix games are zero-sum strategic game involving two players with opposing interests, who can choose a finite number of alternatives (strategies), a game ending score of zero. The goal of game theory to analyze the conflict situation and determine the reasonable behavior of the players during the game, that is. to choose the optimal strategy player (a strategy that ensures the maximum possible secondary gain or minimum possible loss of the middle, with the most adverse action against-nik). Rules of the game are explained through the following example. The game is played by two players, A and B. Player A has at its disposal m alternative, a player B has n alternatives. If player A chooses i-this alternative while player B chooses j-th, then player B pays player A amount aij. Number aij, as the price game, can be positive, negative and zero. If $a_{ij} > 0$, the player receives a certain amount of $V_A(a_i, b_j)$ from player B; if $a_{ij} <0$, then player B receives the amount of $V_B(a_i, b_j)$ from player A, while in the case $a_{ij} = 0$ is no payment. As the player A has at its disposal m alternative, a player B has n alternative, to the game has a total of $m \times n$ possible outcomes and the corresponding amount of payments V. A schematic representation of possible gains A player from player B, depending on the alternative chosen, it can be presented in tabular form, as shown in Table 6-1.

Alternatives of player *A*	**Alternatives of player *B***					
	b_1	b_2	…	b_j	…	b_n
a_1	a_{11}	a_{12}	…	a_{1j}	…	a_{1n}
a_2	a_{21}	a_{22}	…	a_{2j}	…	a_{2n}
⋮	…	…	…	…	…	…
a_i	a_{i1}	a_{i2}	…	a_{ij}	…	a_{in}
⋮	…	…	…	…	…	…
a_m	a_{m1}	a_{m2}	…	a_{mj}	…	a_{mn}

or by using the matrix methods

$$A = \begin{bmatrix} a_{11} & a_{12} & \cdots & a_{1n} \\ \cdots & \cdots & \cdots & \cdots \\ a_{m1} & a_{m2} & \cdots & a_{mn} \end{bmatrix} = \left[a_{ij}\right] \qquad (6.7.1)$$

When the matrix game with a score of zero is filled with a condition

$$V_A(a_i, b_j) + V_B(a_i, b_j) = 0, \qquad (6.7.2)$$

respectively

$V_A(a_i,b_j) = - V_B(a_i,b_j)$. (6.7.3)

Obviously, the first goal of maximizing player functions $V(a_i,b_j)$,, a second player minimization of the same functions.

It should be noted that each player chooses one of the variables ai or bj, which determine the value of the function $V(a_i,b_j)$,; The first player selection strategy, and a second player strategy b_j. The result of the game is uncertain if it knows only one of the strategies. The result of the game is defined only when you know the strategies of both players.

Zero-sum games are a special case of a constant sum games where the sum of the results of both players is constant:

$V_A(a_i,b_j) + V_B(a_i,b_j) = const.$ (6.7.4)

In games with a score zero matrix of payments is entirely determined by the game. For this game, the following rules apply:

- At any time, players behave rationally. Under such circumstances, the most appropriate solution to a game represents the views of the players.
- When choosing your alternative is not a single player has no information on which alternative is chosen his opponent.

Example 6.7.1. The matrix, which is presented in Table 6-2., Ie the gains are shown. Losses two players participating in the game matrix. At stake is the choice of a decision on the strategy of the two opponents which is "Alpha" and another "Beta". It is necessary in a particular case, explain the concept of matrix games with a score of zero.

Matrix game		***Beta***		
		b_1	b_2	b_3
Alpha	a_1	- 1	0	+ 1
	a_2	- 2	- 3	+ 3
	a_3	- 1	+1	- 3

Solution. Competitive strategies are different from those in uncontrolled conditions in that they consciously projected opponent. In the above matrix is assumed that the gains are shown Player "Alpha", represent losses or suspicions that a player "Beta" should pay the first player "Alfa", in all possible combinations of strategies on both sides.

Positive notes correspond to the first player to gain "Alfa" a loss for the other players, "Beta", and vice versa. Eg., If a player "Alpha" a_2 determined strategy, a "Beta" strategy b_3, then the value of the game V= +3, ie. first player "Alpha" will get three points, a second player "Beta" lose the same number of points.

Quantification of Matrix Games

The competition between the two companies, if the gains / losses measures the share of export of goods and services in total sales in a market, you will certainly gain a firm's mark and the loss of a competitive firm. If the gain / loss of custom volume of placements, then the zero sum condition does not apply, unless the total capacity of the market does not change over a long period, and such cases are rare.

In relation to previous allegations should be pointed out that, although game theory presents a mathematical model for determining optimal strategies can be applied and pure qualitative factors. Determining the parameters of matrix games in a qualitative way is inevitable, primarily because it is a social system in which people actively participate. In the following example explains the application of the method kvantafikacije.

Example 6.7. Two of the "Alpha" and "Beta" products are laser printers. The company "Alfa" has developed a new type of printer, where they applied new principles with improved print quality with less toner. For this new principle of production company "Alfa" was granted patent. The management company looks at how to proceed with the new model and consider the following three strategies:

a_1 - plan the production of a large number of new products and cease production of the old model, in order to free up production capacity;

a_2- leave the biggest party in the production of a new model at the same time maintain the operation of the existing line of printers at the previous level;

a_3 - not be released for the production of a new model to the introduction of additional production capacities of theta.

The management of the company "Alpha" believes that the company "Beta" in case its well-informed about the new patent can choose only one of the three reasonable strategy:

b_1 - modernize printers, which currently produces;

b_2 - to keep the production printer that produces, without changes;

b_3 - Try to quickly create their radical new model printer.

It is necessary in this case to explain a method of quantifying a matrix of the game.

Solution. The management of the company "Alpha" compiles its matrix of payment. They determine the size of each intersection of rows and columns of the matrix. This intuitive reasoning are the answer to the question of how the conditions that correspond to var- ious intersections of rows and columns in a matrix ie. different pairs of strategies, may influence the share of the company "Alfa" in the placement of the printer for an extended period. Suppose that after discussion by the leadership of the company "Alpha", agreed upon and developed a matrix of payments, as shown in Table 6-3.

Table 6-3: Qualitative matrix of payment

b_1		**Company *Beta***		
		b_2	b_3	
Company *Alpha*	a_1	Good	Medium	Good
	a_2	Medium	Perfect	Bad
	a_3	Bad	Good	Medium

The evaluation which is shown in Table 6-3. the qualitative character. In order to draft the matrix with quantitative assessments necessary to apply the scale for translating qualitative assessments into quantitative. In Table 6-4. data is linear, with a range of 0 to 6 for translating qualitative assessments into quantitative.

Table 6-4: Range for translating qualitative marks to quantitative marks

Qualitative mark	Quantitative mark
very very bad	0

very bad	1
bad	2
medium	3
good	4
very good	5
perfect	6

After quantification of qualitative given solution, the matrix of the game is shown in Table 6-5.

Table 6-5: Quantitative matrix of payment

b_1		Company "*Beta*"		
		b_2	b_3	
Company "*Alfa*"	a_1	4	3	4
	a_2	3	6	2
	a_3	0	4	3

FREE MATRIX GAMES

The basic question in tasks that are formulated as a matrix game is whether there is an optimal strategy for each player, and how to determine? On example antagonistic matrix game players A and B will explain the procedure for determining optimal strategies for both players and based on that, determine the value (price) of matrix games, keeping in mind the significance of the coefficients a_{ij} of the matrix of payment (6.7.1).

The Lower and Upper Price of Matrix Games

It is assumed that a rational player, which means that each player chooses his strategy so that the outcome of the game is the most favorable for him. A player tries to get as much, and the player B to minimize the loss of the game. This means that they will try to choose the optimal strategy from the set of strategies that are available to them.

Consider a player A first attempt to maximize your winnings. According to the matrix of payment (6.7.1), it has at its disposal an alternative m. If you choose a strategy and with one of the following amounts:

$a_{i1}, a_{i2}, \ldots, a_{in}$

depending on the alternative chosen by the player B. In the worst case, he will receive the smallest of these sums, that is.

$$\min_j a_{ij} = \alpha_i. \qquad (6.8.1)$$

However, choosing a strategy ai, player A has to take into consideration that in this case the player B to choose the optimal strategy bj so that player A receives the amount and $a_{ij} = \alpha_i$.. This means that the player A is the cheapest one strategy - an alternative which fits the biggest gain from the relation (6.8.1), which is

$$\max_i \alpha_i = \max_i \min_j a_{ij} = a_i^*. \qquad (6.8.2)$$

If player A determine its strategy in this way, it represents its lowest guaranteed gain in the case of rational behavior of players B. This means that its gain can not be less than this amount, but may be higher if a player B does not choose its optimal strategy. Therefore, the amount is often referred to as the lower value of the matrix of the game. The strategy provides a gain A player called maksiminimalna strategy (max min strategy), and it is optimal for player A in the case of rational behavior of players B.

Similarly, analyzes efforts player B to minimize his loss. If player B chooses the strategy bj (j-th column of the matrix payments), it pays player A is one of the following amounts:

$a_{1j}, a_{2j}, \ldots, a_{mj}$

depending on the alternative chosen by the player A. In the case of rational behavior of players A player A pays on most of these amounts. Accordingly, the player B is the best strategy, according to which it will have the least amount of payments from the equation

$$\max_i a_{ij} = \beta_j. \quad (6.8.3), \text{ which means}$$

$$\min_j \beta_j = \min_j \max_i a_{ij} = b_j^*. \qquad (6.8.4)$$

If player B determine its strategy in this way, it ensures that they will not pay more than bj *, regardless of the strategy player A. However, his loss may be smaller if the player A does not choose its optimal strategy. Therefore, the amount of bj * called upper value matrix games.

If players choose the optimal strategy according to (6.8.2) and (6.8.4), player A confident that provided the smallest gain since

$$a_i^* = \max_i \min_j a_{ij},$$

and player B is sure not to pay more than

$$b_j^* = \min_j \max_i a_{ij}.$$

It can be shown that for this size still valid inequality:

$$\max_i \min_j a_{ij} £ \min_j \max_i a_{ij}. \qquad (6.8.5)$$

Let us denote the value of matrix games with the bookmark. Then he observes that if both players play their optimal strategy, value matrix games must be between the lower and upper values of matrix games, ie.

$$a_i^* £ v £ b_j^*. \qquad (6.8.6)$$

For some matrix games inequality of the relation (6.8.5) become equations, ie.

$$\max_i \min_j a_{ij} = \min_j \max_i a_{ij} = a_{i^*j^*} = v. \qquad (6.8.7)$$

In these games, it is easy to determine the optimal strategy of player i value matrix games. Element $a_{i^*j^*}$ dies payment that satisfies the equality of the relation (6.8.7) is called a saddle point matrix. Element $a_{i^*j^*}$ represent a saddle point if it meets the requirement that the smallest element of their order but at the same time the largest component of its column.

The importance of a saddle point matrix of payments is as follows:

- It determines the optimal strategies for both players. Red matrix payments with index i* represents the optimal strategy for player A, while columns of payments with the index j* determines the optimal strategy for Player B. For matrix game in which both players play always the same strategy we say that a game with clear strategy, or that they this free matrix games.

- The value of the element that represents a saddle point is both a value matrix game can be seen from the relation (6.8.7).

Example 6.8.1. The game involved two players A and B. Both players have available three alternatives. Alternative A player given by rows and by columns B players.

Elements of matrices aij payments indicate the amounts that player B pays player A after playing certain alternatives. Let's play the following data matrix payments

$$\begin{bmatrix} -4 & 4 & -1 \\ 3 & 3 & 2 \\ 8 & -2 & 0 \end{bmatrix}$$

Solution. If a player chooses A first strategy of its smallest possible gain is -4 N.J. (this represents player A's loss). If you choose the second strategy, the least you can get is 2 N.J., and if you choose the third lowest possible payout will amount to -2. Therefore he will not reason like this: it is best to choose a different strategy, because we will in any case get at least 2 N.J., and if the player B does not choose for themselves the best strategy may be that we get and 3 N.J.

If player B chooses first strategy most that can lose 8 N.J., if selected another 4 N.J., and in the selection of the third 2 N.J. Therefore, for the player B is best to choose the third strategy because it will not lose more than 2 N.J., regardless of the strategy chosen player A.

The process of determining the strategy of players A consists in to determine the smallest elements aij for each row, and then the smallest of all the elements in the highest ranks of the elect. Therefore, in addition to the column matrix of payment, determine at least the elements aij for each row. These are: -4 for the first row, 2 for second and third row of -2. The largest of these elements is 2, that is. holds that:

$$\max_i \min_j a_{ij} = \max_i \{-4, 2, -2\} = \max_i \{a_{11}, a_{23}, a_{32}\} = a_{23} = a_2^* = 2.$$

In determining the optimal strategy, player B choosing the biggest elements in each column, and the largest of all these elements we choose the smallest. In the row below the matrix of payment, determine the greatest elements aij for each column: 8 for first, 4 second and 2 in the third column. The smallest of all these elements is 2, that is. okay

$$\min_j \max_i a_{ij} = \min_j \{8, 4, 2\} = \min_j \{a_{31}, a_{21}, a_{23}\} = a_{23} = b_3^* = 2.$$

The described method can also be represented as follows:

$$\begin{array}{c} \min_j a_{ij} \\ \begin{bmatrix} -4 & 4 & -1 \\ 3 & 3 & 2 \\ 8 & -2 & 0 \end{bmatrix} \begin{matrix} -4 \\ 2 \\ -2 \end{matrix} \\ \max_i a_{ij} \end{array}$$

Now, the following is valid:

$$\max_i \min_j a_{ij} = a_{23} = \min_j \max_i a_{ij} = a_{23} = 2,$$

This means that the matrix has a saddle point of payment, which is located at the intersection of other alternatives for players and A third alternative for the players B. Saddle point represents the value of the game, in our case, $a_{23} = 2$. The value of the game is the smallest possible gain for player A, if you choose the optimum strategy and maximum possible loss for the player B, if he chooses the optimal strategy. This further means that the game has a clear strategy: Player A is best to always chooses a different strategy, because it provides a gain of at least 2 of the monetary unit; Player B is the third best strategy because it ensures that they will not in any case paying more than 2 pennies. The value of the game is equal to the value of the element that represents a saddle point, ie. $v = a_{23}$ n = 2.

Example 6.8.2. After the conflict situation is a relationship of buyer and seller of goods. The customer has alternatives Kl, K2, K3 and K4 (eg. Alternative methods of payment can be a buyer in cash, bank card, check, credit, and similar), a vendor with alternatives P1, P2, P3 and P4 (for example, time limits payment options: 10, 15, 20, 30 days). Depending on the payment method the customer and the payment deadline, the seller approves credit note given in% as shown in the following table. It is necessary to find the optimal strategy for the buyer and seller, and determine how allowances will be for the buyer if the buyer and the seller choose the optimal strategy.

	Seller			
Buyer	P_1	P_2	P_3	P_4
K_1	6	6	5	7
K_2	9	5	4	8
K_3	8	7	6	9
K_4	7	3	5	7

Solution. First check whether there is a saddle point in the matrix of payment. This is done by using Wald's pessimistic criteria shown in the matrix form.

$$\begin{array}{c c} \begin{bmatrix} 6 & 6 & 5 & 5 \\ 9 & 5 & 4 & 8 \\ 8 & 7 & 6 & 9 \\ 7 & 3 & 5 & 7 \end{bmatrix} & \left.\begin{matrix} 5 \\ 4 \\ 6 \\ 3 \end{matrix}\right\} 6 \\ \hline \underbrace{\begin{matrix} 9 & 7 & 6 & 8 \end{matrix}}_{6} & \end{array}$$

The following equation holds:

$$\max_i \min_j a_{ij} = a_{33} = \min_j \max_i a_{ij} = a_{33} = 6,$$

means that there is a saddle point, but it is a matrix with pure strategy games. The optimal strategy for the customer is K3 and the seller is P3. If the buyer and the seller choose their optimal strategy, then the smallest possible allowances for customer amount to 6%, and for the seller, if he chooses his optimal strategy, the highest possible allowances are 6%.

CHAPTER 7

WATER NETWORK DESIGN USING A MULTIOBJECTIVE REAL OPTIONS FRAMEWORK

João Marques,[1] Maria Cunha,[1] Dragan Savić,[2] and Orazio Giustolisi[3]

[1]Marine and Environmental Sciences Centre (MARE), Department of Civil Engineering, University of Coimbra, Coimbra, Portugal

[2]Centre for Water Systems, School of Engineering, Computing and Mathematics, University of Exeter, Exeter, UK

[3]Department of Civil Engineering and Architecture, Technical University of Bari, Bari, Italy

ABSTRACT

Water distribution networks (WDNs) are an essential element of urban infrastructure. To achieve a good level of performance, the traditional design of WDNs based on expected future conditions should be replaced by a flexible design, using real options (ROs), that accounts for uncertainty by taking a broader view of possible future options. This work proposes a multiobjective ROs framework that sets out to reduce costs, minimize hydraulic pressure deficiency, and a third objective for minimizing carbon emissions. A multiobjective simulated annealing algorithm is used to identify the Pareto-optimal solutions, thus enabling a trade-off analysis between solutions. These trade-offs show that a low pressure deficit solution is achieved by increasing investment at a much faster rate after a certain pressure deficit threshold (60 m). Also, the pressure deficits can only be reduced by increasing carbon emissions. Finally, this work also emphasizes the importance of including carbon emissions as a specific objective by comparing the results of the proposed model and another one that did not cover the environmental objective. The results show that it is possible to reduce CO_2 for the same level of capital expenditure or the same level of network pressure deficits if carbon emissions are minimized in the optimization process.

INTRODUCTION

Water distribution networks (WDNs) are an important element of urban infrastructure whose purpose is to deliver water of appropriate quality and quantity to users without major interruptions. People normally have high expectations regarding the level of service provided by water utilities. Failure should be prevented at all costs, and so water companies wish to maintain high performance and quality standards. The proper functioning of these infrastructure elements under normal and abnormal operation conditions can only be guaranteed if uncertainty is proactively taken into account at the planning stage. In the traditional design of WDNs, a fixed design is determined by assuming a deterministic projected future demand. This can lead to the underdesign or overdesign of WDNs. With this shortcoming in mind, the paradigm has to be changed by using real options (ROs) to achieve a flexible design.

Myers [1] was the first to propose ROs and mention using option pricing theory to value nonfinancial or "real" investments with learning and flexibility. A number of studies have been published where the ROs concepts have been applied in various fields, such as industrial processes [2], energy systems [3], mining projects [4], flooding problems [5], and maritime costal defences [6]. For the design of WDNs, ROs have been explored by Huang et al. [7], who use ROs accounting for future demand uncertainty in a single-objective approach. Basupi and Kapelan [8, 9], also considering demand uncertainty, propose a flexible design using a multiobjective optimization model with two objectives, cost minimization and the maximization of end resilience. The intervention plans set out in these papers [7–9] are defined in stages but they do not allow for different expansion areas for the WDN during the planning horizon. Here we aim to introduce a different analysis by proposing a ROs approach that considers expansion scenarios of the network at different time phases and by taking into account environmental concerns using a multiobjective optimization model with three objectives. In Marques et al. [10], the carbon emissions were included in the costs objective. This is not satisfactory as carbon prices are very volatile and the consideration of a fixed value for the entire planning horizon of the network can lead to the impact of carbon emissions being misrepresented. Furthermore, there are aspects of pipe manufacturing and energy production where, in some cases, the carbon emissions related to the industrial process are already included in the final price of the product and thus should not be duplicated in the model. To overcome the shortcomings of past works in which carbon emissions are monetized and minimized within a cost minimization function [10], we propose to evaluate the impact of carbon emissions in a separate objective. Additionally, some comparisons are made to identify the advantages of considering these environmental concerns in the ROs approach.

The initial attempts to achieve optimal design and operation of WDNs focused on a single-objective strategy, that is, solving a least-cost optimization problem whose primary decision variables were the pipe diameters. One of the first attempts was made by Alperovits and Shamir [11]. However, a number of investigators have noted that, from a technical standpoint, the design of WDNs could be improved by using a multiobjective strategy. That approach allows the best trade-offs to be identified among different objectives, such as cost, reliability and environmental issues. As explained by Savic [12], single-objective optimization approaches have a number of limitations and a multiobjective technique is proposed to avoid these difficulties. Therefore, this work presents a multiobjective optimization model that includes the traditional objectives of minimizing costs and improving hydraulic performance, along with a third objective in the optimization process to explicitly take into account environmental concerns by minimizing the carbon emissions. Some studies even include environmental impacts in the water infrastructure planning, e.g., Herstein et al. [13], who present an index-based method to assess the environmental impact of WDN and D'Ercole et al. [14] who optimize the WDN performance by taking environmental impacts into account. However, there is a gap in the literature with respect to handling environmental issues in a multiphase design scheme through the use of real options (ROs) to achieve a flexible design. Our work intends to fill this gap.

This work presents an optimization method for the design and operation of WDNs that implements a multiphase strategy to achieve a flexible design by taking uncertainty into consideration when planning short- to long-term investments. The multiphase design of WDNs can be defined as design that is carried out in phases using small time horizons at each phase (rather than considering the full planning horizon). This enables decision makers to actively manage the configuration of the network and easily adapt it to new situations. A multiphase design is more cost effective for a long planning horizon than a single-phase design because the construction costs are incurred progressively over time and not all at once (year zero of the planning horizon), which is the case for a single phase. Deferring costs saves money. Furthermore, multiphase design makes it possible to adjust the WDN's capacity if conditions turn out to be different from those projected. The traditional single-phase solutions, however, can result in a system that may need upgrading from time to time or require excessive initial investment in capacity that may not be required for a long time.

The rest of the paper is organized as follows: Section 2 sets out the multiobjective optimization algorithm, followed in Section 3 by the multiobjective decision model established according to the ROs approach. Then a detailed description of the case study is presented in Section 4, including the future scenarios options, after which the results are given in Section 5 and some comparisons drawn. Section 6 lists some conclusions.

MULTIOBJECTIVE SIMULATED ANNEALING

The multiobjective optimization problem is solved by a simulated annealing heuristic [15]. Simulated annealing is a popular search algorithm based on an analogy with the physical process of cooling a material in a heat bath [16]. Annealing is the process of heating a solid until its melting temperature and then slowly reducing the temperature. If the cooling process is sufficiently slow the process forms a new material with a well-organized structure and the lowest internal energy state. The notion of slowly decreasing the temperature is adapted in the simulated annealing algorithm to guide the optimization process by slowly reducing the probability of accepting worse solutions (those that have worse values of the objectives) during the exploration of the solution space. The possibility of accepting worse solutions is a crucial property that allows the algorithm to escape from local optimums and enables a more widespread search. The algorithm comes to an end when the temperature parameter achieves the desired level.

A literature review shows that simulated annealing has been used in various fields, including regional wastewater system planning by Zeferino et al. [17] and groundwater monitoring networks by Nunes et al. [18], where it performed well. Cunha and Sousa [19] and Reca et al. [20, 21] have demonstrated the successful use of simulated annealing in WDNs for single-objective optimization models. In a previous work by Marques et al. [10], a simulated annealing algorithm was proposed and used in a two-objective optimization model for the design of WDNs. A three-objective optimization model (the type of model proposed here) is more complex than a two-objective model, because the search for the Pareto-optimal front becomes harder due to the increase of the number of incomparable solutions and also because it is more difficult to classify the dominance relationship between solutions to guide the search.

The multiobjective simulated annealing algorithm is used in a three-objective model to search for the Pareto-optimal front by generating potential solutions in the solution space and evaluating the domination status between new solutions and solutions that have already been found. The domination status is represented schematically by Figure 1 between two solutions “a” and “b” and is computed by multiplying the change in values (if this is a nonzero value) of the objectives ($O1$, $O2$, and $O3$). In a three-objective optimization problem, the domination is given by a parallelepiped volume, as shown in Figure 1, where the lengths of the edges are given by the change in values of the three objectives. If these differences are large, the volume or dominance between solutions is high. But if the volume is small, so is the dominance of the solutions.

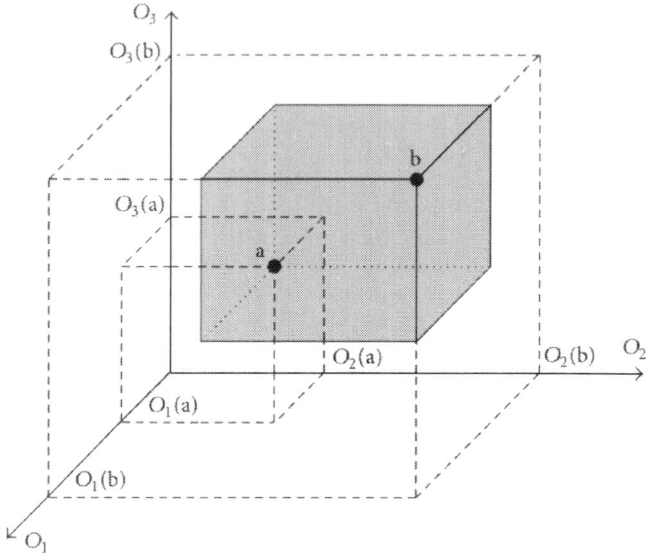

Figure 1: Dominance between solutions "a" and "b."

Depending on the domination status, the potential solutions are accepted or rejected according to the Metropolis criterion [16], which guides the optimization process until the stop criterion is met. In addition, the ROs approach divides the planning horizon into design phases where different decisions are taken. These multiple design phases increase the complexity of the problem. Furthermore, to easily explore the solutions provided by the optimization model, a visualization tool AEROVIS [22] is used to plot the Pareto solutions.

MULTIOBJECTIVE OPTIMIZATION MODEL FORMULATION

Decision Variables

The approach described here aims to define the flexible design for WDNs by achieving the minimization of three objectives: investment and operation costs, pressure deficit, and carbon emissions. The decision variables are the commercial pipe diameters of the WDN. The flexible design strategy determines the value of the decision variables for the first phase and also for all the future decision options that arise from all the possible decision paths (corresponding to different scenarios of expansion of the network for different time periods). However, the values of the decision variables that are determined for the first phase are effectively the ones that are required now, and therefore it is very important to ensure that these decisions are flexible enough to cope with the network conditions in the next phases. The values of the decision variables for the future time periods (phases) can be adjusted as new information arises.

Objective Functions

We formulated the multiobjective optimization model to minimize costs, pressure deficits, and carbon emissions. These three objectives are minimized simultaneously to allow identifying the major trade-offs between them and are represented in the following expressions:

$$O1 = \min\left(\mathrm{Ci} + \mathrm{Cf}\right) \quad (1)$$

$$O2 = \min\left(\mathrm{HPD}\right) \quad (2)$$

$$O3 = \min\left(\mathrm{tonCO_2}\right), \quad (3)$$

where Ci is cost of the initial solution to be implemented in the first period (USD); Cf are future costs (USD); HPD are hydraulic pressure deficits (m); and ton CO_2 are tonnes of CO_2 emitted (ton).

Equation (1), representing the first objective, is given as the sum of the initial solution costs to be implemented in the first period or phase and the future costs for the planning horizon. The term Ci computes the cost for the first period and is calculated by

$$\mathrm{Ci} = \sum_{i=1}^{\mathrm{NPI}} \left(\mathrm{Cpipe}_{i,1}\right) + \sum_{j=1}^{\mathrm{NPU}} \left(\mathrm{Cps}_{j,1}\right) + \sum_{d=1}^{\mathrm{NDC}} \mathrm{Ce}_{d,1}, \quad (4)$$

where NPI is number of pipes in the network; $\mathrm{Cpipe}_{i,1}$ is cost of pipe i in period t=1; NPU is number of pumps in the network; Cps$j$,1 is pumping station costs of pump j in the period t=1; NDC is number of demand conditions considered for the design; and Ced,1 is present value cost of energy (computed for the time period) under demand condition d in period t=1.

Summing the cost of pipes, the cost of pumps, and the present value of the energy cost we get the total cost for the solution of t=1. The other term of the objective function $O1$ in (1) represents the future cost of pipes, pumps, and energy of all the scenarios. These costs are weighted by the corresponding probability of each scenario of expansion as showed in

$$\mathrm{Cf} = \sum_{s=1}^{\mathrm{NS}} \sum_{t=2}^{\mathrm{NTI}} \left(\mathrm{Cfuture}_{t,s} \cdot \prod_{\mathrm{nt}=2}^{t} \mathrm{prob}_{\mathrm{nt},s} \right), \tag{5}$$

where NS is number of scenarios; NTI is number of periods into which the planning horizon is divided; $\mathrm{Cfuture}_{t,s}$ is cost of future designs in scenario s for period t; and $\mathrm{probn}_{t,s}$ is probability of $s$in period nt (nt represents the number of time periods, from the second to the one under analysis).

The future costs are computed, for each path, by summing all costs incurred in each period, starting from the second time period. In each future option the costs are computed by multiplying the cost of each design option by the probability of taking that option. The term $\mathrm{Cfuture}_{t,}$ is computed in (6), for all future options established after t=1 (the costs for the first period are already calculated by Ci).

$$\mathrm{Cfuture}_{t,s} = \left[\sum_{i=1}^{\mathrm{NPI}} \mathrm{Cpipe}_{i,t,s} + \sum_{j=1}^{\mathrm{NPU}} \left(\mathrm{Cps}_{j,t,s} \right) + \sum_{d=1}^{\mathrm{NDC}} \left(\mathrm{Ce}_{d,t,s} \right) \right] \cdot \frac{1}{(1+\mathrm{IR})^{Y_t}}, \quad \forall t \in \mathrm{NTI};\ \forall s \in \mathrm{NS}, \tag{6}$$

where $\mathrm{Cpipe}_{i,,s}$ is cost of pipe i installed in period t and in scenario s; $\mathrm{Cps}_{j,t,s}$ is pumping station costs of pump j installed in period t and in scenario s; $\mathrm{Ce}_{d,t,s}$ is present value cost of energy (updated for the first year of the time period t) under demand condition d for period t and in scenario s; IR is annual interest rate for updating cost; and Yt is year when costs will be incurred for period t.

The future costs are computed by a sum of 3 terms: the first term computes the cost of pipes, the second term computes the cost of installing pumps, and the last term computes the energy cost of pump operation in the time period. The present value of these costs is then computed for the year when the costs will be incurred. Summing the terms Ci with Cf in (1), the total cost for the whole planning horizon is determined. The uncertainty of future network expansion is taken into account, according to a decision tree, where each path has its own probability.

The second objective function ($O2$), given in (2), aims to minimize the total pressure deficits, considering all possible future options that can be taken for the planning horizon. This is computed by summing the node pressure deficits for the different possible scenarios according to

$$\mathrm{HPD} = \sum_{s=1}^{\mathrm{NS}} \sum_{t=2}^{\mathrm{NTI}} \sum_{d=1}^{\mathrm{NDC}} \sum_{n=1}^{\mathrm{NN}} \max\left\{0; \left(\mathrm{Pdm}_{n,d} - P_{n,d,t,s}\right)\right\}, \tag{7}$$

where NN is number of nodes; $\mathrm{Pdm}_{n,d}$ is minimum desirable pressure at node n for demand condition d; and $P_{n,d,t,s}$ is pressure at node n at demand condition d for time period t and in scenario s.

Equation (7) computes the pressure deficits for all scenarios, all time periods, after$t$=1, for all the demand conditions, and for all the nodes. For t=1, nodal pressures have to be totally satisfied (i.e., nodal pressures equal to or higher than minimum desirable pressures). This sum can be used as a measure of the network performance over the entire planning horizon.

Finally, the third objective ($O3$) in (3) includes the minimization of carbon emissions due to installation and operation of the WDN and is defined in (8). The carbon emissions are evaluated by considering the whole life cycle, including the extraction of raw materials, transport, manufacture, assembly, installation, disassembly, demolition and/or decomposition [23]. The carbon emissions from the energy used during the network's operation, mainly by pumping stations, are also computed. We do not consider the other elements of the network, for simplification purposes.

$$\text{tonCO}_2 = \sum_{i=1}^{\text{NPI}} (\text{CO}_2\text{pipe}_{i,1}) + \sum_{d=1}^{\text{NDC}} (\text{CO}_2\text{Ce}_{d,1})$$

$$+ \sum_{s=1}^{\text{NS}} \sum_{t=2}^{\text{NTI}} \left\{ \left[\sum_{i=1}^{\text{NPI}} (\text{CO}_2\text{pipe}_{i,t,s}) + \sum_{d=1}^{\text{NDC}} (\text{CO}_2\text{Ce}_{d,t,s}) \right] \right.$$

$$\left. \cdot \prod_{\text{nt}=2}^{t} \text{prob}_{\text{nt},s} \right\}, \tag{8}$$

where $CO_2pipe_{i,t,s}$ is carbon emissions of pipe i installed in period t and in scenario s and $CO_2Ce_{d,t,s}$ are carbon emissions of energy consumed by pump operation under demand condition d for period t and in scenario s.

The carbon costs are given by summing the emissions related to pipeline construction with the carbon emissions by pump operation. These emissions are computed for the first time period or phase by the first two terms in (8) and by a weighted sum of the carbon emissions of all the possible future scenarios that can be taken. The future carbon emissions are also computed by summing the pipeline construction with the pump operation carbon emissions.

Constraints

The multiobjective model also includes a set of constraints. The minimum admissible pressures are used to compute the lowest value of pressure that can be reached according to the probabilities of scenarios and calculated by the following expression:

$$P_{n,d,t,s} \geq \left[(\text{Pdm}_{n,d} - \text{Pam}_{n,d}) \cdot \prod_{\text{nt}=2}^{t} \text{prob}_{\text{nt},s} \right] + \text{Pam}_{n,d}$$

$$\forall n \in \text{NN};\ \forall d \in \text{NDC};\ \forall t \in \text{NTI};\ \forall s \in \text{NS}, \tag{9}$$

where $Pam_{n,d}$ is minimum admissible pressure at node n for demand condition d.

The constraint of (9) is intended to obtain higher minimum pressures, and thus fewer pressure violations, for scenarios with high probabilities of occurrence. This can be seen as a procedure to increase the performance of the network to handle certain, more probable, situations. The model also includes other constraints to verify nodal continuity (see (10)), to compute the head loss of the pipes (see (11)), and to guarantee a minimum pipe diameter (see (12)) for each pipe, the use of a set of commercial diameters (see (13)) and the assignment of only one commercial diameter for each pipe (see (14)).

$$\sum_{i=1}^{NPI} I_{n,i} Q_{i,d,t,s} = QC_{n,d,t,s}$$

$$\forall n \in NN; \ \forall d \in NDC; \ \forall t \in NTI; \ \forall s \in NS \qquad (10)$$

$$\Delta H_{i,d,t,s} = K_i Q^{\alpha}_{i,d,t,s}$$

$$\forall n \in NN; \ \forall d \in NDC; \ \forall t \in NTI; \ \forall s \in NS \qquad (11)$$

$$D_i \geq Dmin_i \quad \forall i \in NPI \qquad (12)$$

$$D_i = \sum_{d=1}^{ND} YD_{d,i} \cdot Dcom_{d,i} \quad \forall i \in NPI \qquad (13)$$

$$\sum_{d=1}^{ND} YD_{d,i} = 1 \quad \forall i \in NPI, \qquad (14)$$

where $I_{n,i}$ is incidence matrix of the network; $Q_{i,d,t,s}$ is flow in pipe i under demand condition d for period t and scenario s (m^3 /s); $QC_{n,d,t,s}$ is consumption at node n under demand condition d for period t and scenario s (m^3 /s); NN is number of nodes; $\Delta H_{i,d,t,s}$ is head loss in pipe i under demand condition d for period t and scenario s; Ki, α are coefficients that depend on the physical characteristics of pipe *i*; D_i is diameter of pipe *i*; $Dmin_i$ is minimum diameter of pipe i; $YD_{d,i}$ is binary variable to represent the use of diameter d in pipe i; $Dcom_{d,i}$ is commercial diameter d assigned to pipe i; and ND is number of commercial diameters.

The optimization model was linked to EPANETpdd [24] pressure-driven hydraulic simulator to verify the hydraulic constraints. This is an extension of the EPANET [25] demand-driven simulator. Water distribution simulation models are usually divided into two groups, the demand-driven and the pressure-driven models. Demand-driven is the traditional analysis and assumes that nodal demands are known and are supplied independently of the pressure values. The pressure-driven analysis is based on a more recent modelling strategy that considers demand not fixed but depending on the pressures in the network [26, 27]. Pressure-driven analysis allows a more realistic prediction of the hydraulic system behavior of WDN [27, 28]. From a computational point of view, demand-driven analysis is more efficient than pressure-driven due to the increased complexity of the hydraulic modelling problem as set out by Giustolisi and Walski [28]. However, pressure-driven analysis is required here to predict the hydraulic behavior of the network in pressure-deficient conditions. EPANETpdd is used to calculate the hydraulic equations by an iterative procedure presented by Morley and Tricarico [24]. This procedure follows the head-flow relationship proposed by Wagner et al. [29].

CASE STUDY

We use a WDN adapted from Walski et al. [30] to demonstrate the applicability of our approach, and this is presented in Figure 2. It is a small network supplied by three reservoirs with fixed levels and a pumping station at link 1 to transfer energy to the flow from reservoir 1.

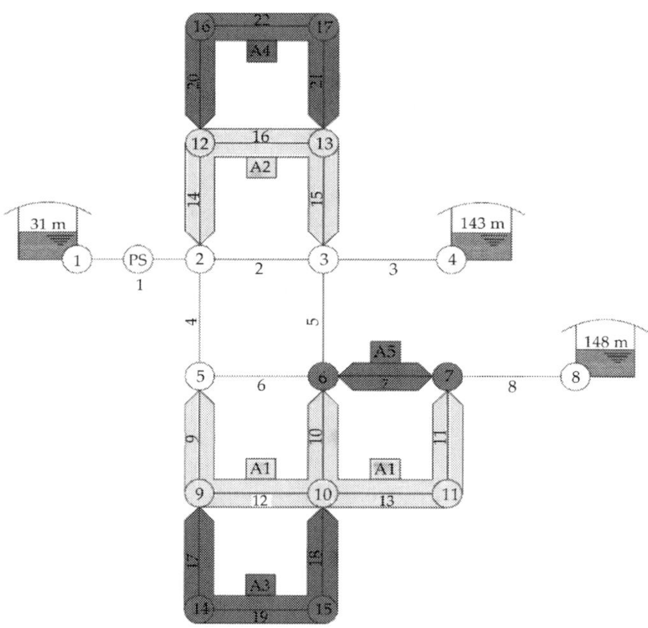

Figure 2: Water distribution network.

The grey areas (A1 to A5) of the network in Figure 2 are explained at the end of this section. The grey shades are used to identify the network areas that will be expanded at different phases of the planning horizon or with a population shrinkage. Light grey represents the network areas to be installed in near future (A1, A2) and dark grey represents the areas that will be installed over the long term (A3, A4) and A5 where a population shrinkage is considered.

The characteristics of the nodes including the areas, the ground elevation, the demand for conditions (1) and (2), and two different minimum pressures for each demand condition are given in Table 1. We consider two different minimum pressure values to design the network for the long term, the desired pressure (minimum pressure to provide a good level of network service) and admissible pressure (minimum allowed pressures in the network below which no demand can be delivered). Pressure at the nodes can be lower than the desired pressure but not lower than the admissible pressure.

Table 1: Characteristics of the nodes

Node	Area	Ground elevation (m)	Nodal demand condition (l/s)		Minimum desirable pressure (m)		Minimum admissible pressure (m)	
			(1)	(2)	(1)	(2)	(1)	(2)
1		36.5	Reservoir at the level of 30.5 m					
2		30.5	0	0	28.1	17.6	21.1	10.6
3		106.7	31.6	47.3	28.1	17.6	21.1	10.6
4		117.4	Reservoir at the level of 143.3 m					
5		106.7	31.6	47.3	28.1	17.6	21.1	10.6
6	A5	106.7	126.2	189.3	28.1	17.6	21.1	10.6
7	A5	106.7	63.1	94.6	28.1	17.6	21.1	10.6
8		121.9	Reservoir at the level of 147.8 m					
9	A1	106.7	31.6	47.3	28.1	17.6	21.1	10.6
10	A1	106.7	31.6	47.3	28.1	17.6	21.1	10.6
11	A1	106.7	31.6	47.3	28.1	17.6	21.1	10.6
12	A2	106.7	31.6	47.3	28.1	17.6	21.1	10.6
13	A2	106.7	31.6	47.3	28.1	17.6	21.1	10.6
14	A3	106.7	31.6	47.3	28.1	17.6	21.1	10.6
15	A3	106.7	31.6	47.3	28.1	17.6	21.1	10.6

16	A4	106.7	31.6	47.3	28.1	17.6	21.1	10.6
17	A4	106.7	31.6	47.3	28.1	17.6	21.1	10.6

The characteristics of the pipes are given in Table 2, namely, the initial and final nodes, the length, and the pipes to be installed according to the areas of Figure 2.

Table 2: Characteristics of the pipes

Pipe ID	Initial node	Final node	Length (m)	Area
1	1	2	Pump	
2	2	3	3218.7	
3	3	4	3218.7	
4	2	5	1609.3	
5	3	6	1609.3	
6	5	6	3218.7	
7	6	7	3218.7	A5
8	7	8	1609.3	
9	5	9	1609.3	A1
10	6	10	1609.3	A1
11	7	11	1609.3	A1
12	9	10	3218.7	A1
13	10	11	3218.7	A1
14	2	12	1609.3	A2
15	3	13	1609.3	A2
16	12	13	3218.7	A2
17	9	14	1609.3	A3
18	10	15	1609.3	A3
19	14	15	3218.7	A3
20	12	16	1609.3	A4
21	13	17	1609.3	A4
22	16	17	3218.7	A4

In relation to the pump (pipe ID 1), the efficiency is 75% and the daily consumption is 20 hours at demand condition (1) and the other 4 hours at demand condition (2). The energy costs are 0.075 USD/kWh and should be evaluated for a 60-year period at a discount rate of 4% per year. This rate was fixed based on recommendations by Wu et al. [31].

Eight possible commercial diameters are considered for the optimal design of the WDN and these are presented in Table 3, including (for each commercial diameter) the unit cost, the Hazen-Williams coefficients, and the carbon emissions. These carbon emissions are computed by the procedure described by Marques et al. [23].

Table 3: Diameter, unit costs, carbon emissions, and Hazen-Williams coefficients

Diameters (mm)	**Unit costs (USD/m)**	**Hazen-Williams coefficients**	**Carbon emissions (ton CO_2/m)**
152.4	49.5	100	0.5
203.2	63.3	100	0.6
254	94.8	100	0.7
304.8	132.9	100	0.8
355.6	170.9	100	0.9
406.4	194.9	100	1.0
457.2	225.1	100	1.1
508	262.8	100	1.1

The planning horizon for this case study is set at 60 years and divided into three periods or phases of 20 years each, when different options can be taken. These options are stated according to the uncertainty of future outcomes and are organized by ROs in terms of possible future scenarios. The dynamics of urban growth over the lifetime horizon have a considerable impact on the WDN and should therefore be taken into account at the infrastructure planning stage.

A decision tree is built for the case study using the various decision paths that can be taken in future and it is presented in Figure 3.

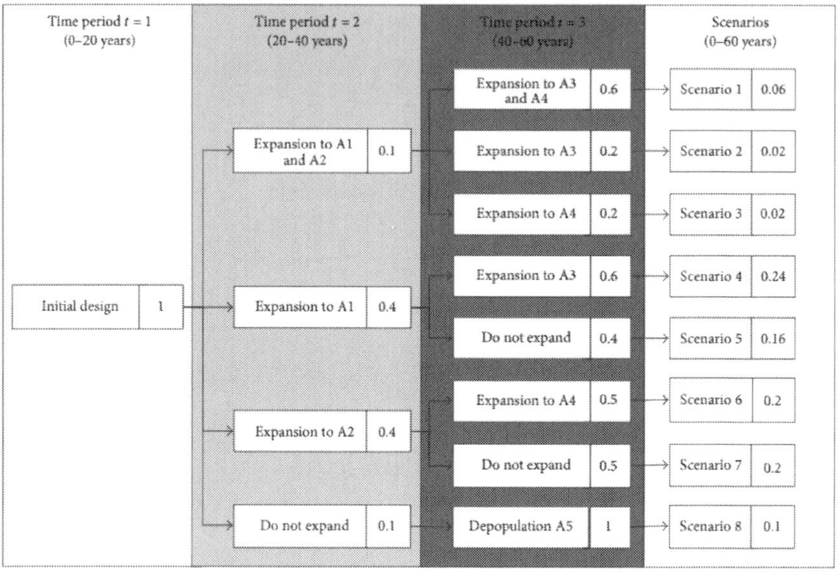

Figure 3: Decision tree for the planning horizon and occurrence probability.

The planning horizon, represented in the decision tree, is divided into three time periods. In the first time period, t=1, the network design is required for pipes 1 to 8, allowing for future uncertainty. In t=2 (light grey), there are four options available, expansion to A1 and A2, expansion to A1, expansion to A2 and do not expand. For the last time period (dark grey), t=3, other options are possible, that is, expansion to A3 and A4, expansion to A3, expansion to A4, and do not expand. A population shrinkage is also considered, A5, that sees consumption fall by 30%. These future conditions are considered the most probable and are organized in the decision tree by phased interventions in the network that can be made by following a decision path in the decision tree. Each decision option has a probability shown in the squares in Figure 3 and the probabilities of scenarios given in the last branches of the decision tree are computed with these values by multiplying all the probability values on the decision path of the scenario. In real world applications these probabilities can be given by expert judgment.

RESULTS

Keeping the problem formulation and the case study in mind, the results are identified for all the design phases and for all scenarios of the planning horizon. However, as previously stated, just the first phase of the design has to be built now and the process can be reassessed in subsequent phases to find more accurate solutions once some uncertainty has been eliminated. The solutions were obtained by using the multiobjective simulated annealing algorithm and the results are plotted in Figure 4.

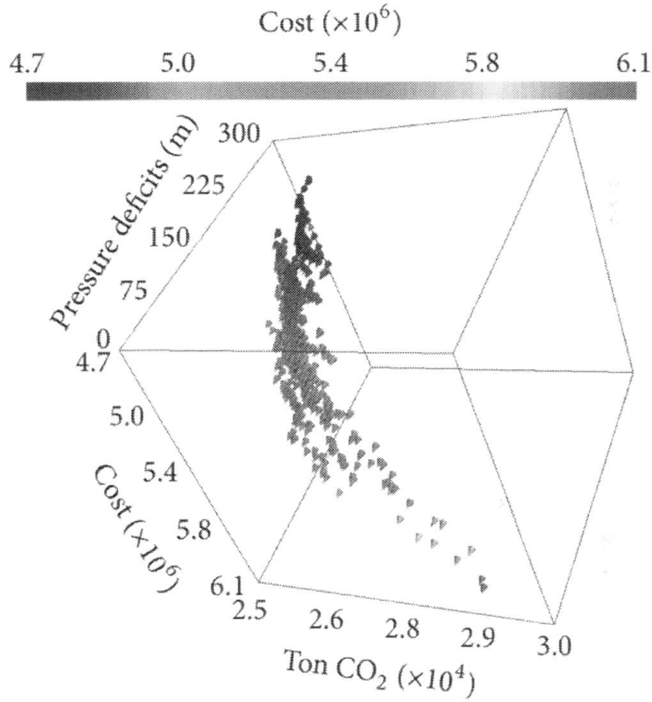

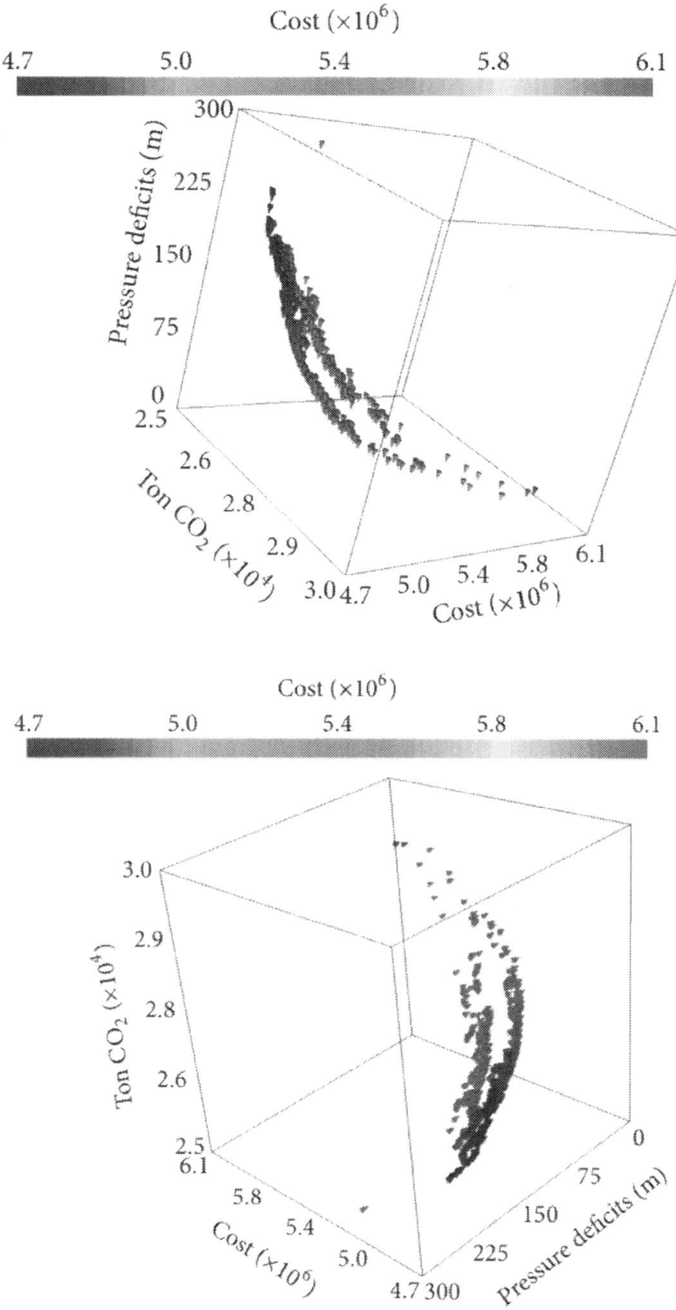

Figure 4: Pareto front from different perspectives.

A global view of the three objectives' possible Pareto front is shown in Figure 4 by a set of 600 solutions. This representation makes use of the AEROVIS visualization tool developed by Kollat and Reed [32] and Kollat et al. [22]. These solutions are plotted for different perspectives to try to facilitate the visualization of the shape of the Pareto front surface so as to indicate the type of trade-off between the different objectives.

The color scheme of the figures shows the cost of the solutions in US dollars. The colder (blue) shades represent the lower values of cost and the warmer (red) shades represent the higher values. The color scheme and the relevant cost values are indicated in the color bars above the figures.

The trade-offs between the multiobjective optimization solutions are explored in detail by plotting these results in two-dimensional graphs by pair of objectives. This is done using the AEROVIS visualization tool to emphasize the nondominated solutions between each pair of objectives and soften the others. Figure 5 shows the costs plotted with pressure deficits.

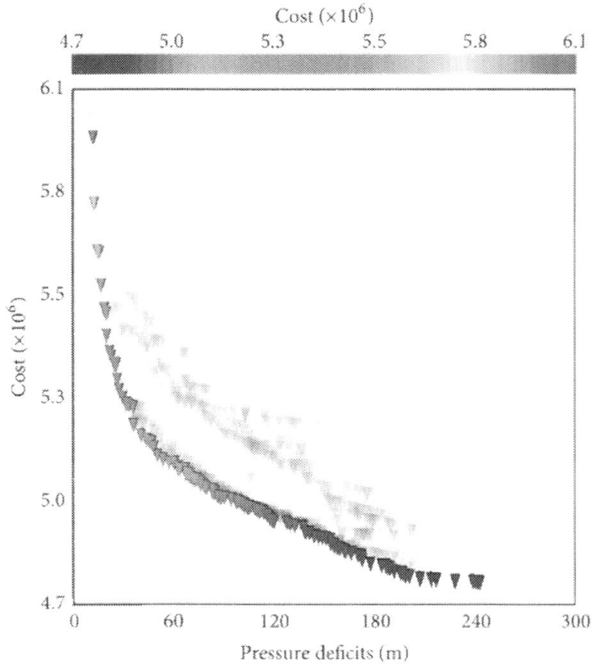

Figure 5: Pareto front of cost versus pressure deficit.

This figure shows that increasing the cost leads to decreasing the hydraulic pressure deficits. For total pressures deficits below approximately 60 (m), the rate of cost increase is much higher than for pressure deficits above 60 (m). In relation to cost and carbon emissions, the trade-offs between these two objectives are presented in Figure 6.

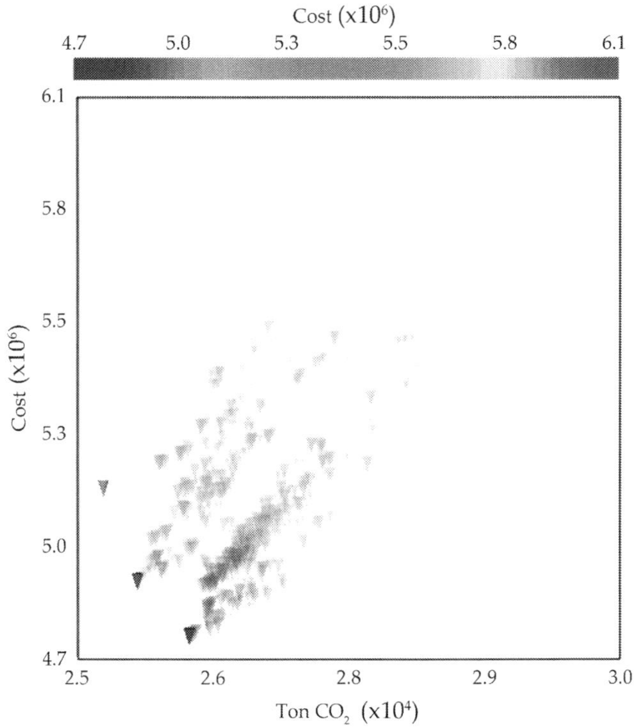

Figure 6: Pareto front of cost versus ton CO_2.

It can be seen from the figure that the carbon emissions are not in conflict with the cost objective. Roughly, decreasing the amount of material for pipes and the energy consumed by pumps makes it possible to reduce both the cost and the carbon emissions. Finally, Figure 7 shows the relation between the carbon emissions and the pressure deficits.

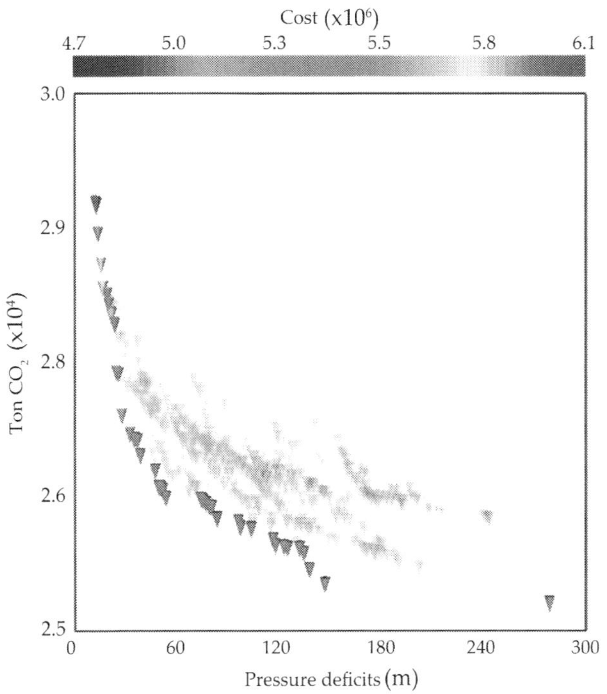

Figure 7: Pareto front of ton CO_2 versus pressure deficits.

It can be seen from this figure that the hydraulic pressure deficit is reduced by increasing the carbon emissions. In fact, the reduction of the pressure deficits is achieved by increasing the network's hydraulic capacity. This can be done by introducing large-diameter pipe sections or by increasing the energy used for pumping. Both imply more carbon emissions through the rise in the quantity of pipe material needed and in the energy consumed by pumps.

To understand the importance of including a carbon emissions objective in the optimization model, the results of the optimization model described previously, abbreviated as OM1, are compared with the results of the optimization model OM2 that do not include carbon emissions as an objective to minimize. OM1 (described in Section 3) has three objective functions included in expressions (1), (2), and (3), to minimize costs, pressure deficits, and carbon emissions, respectively. The OM2 model has the two objective functions presented in expressions (15). These expressions are equal to expressions (1) and (2) of OM1.

$$O1 = \min(Ci + Cf)$$

$$O2 = \min(HPD). \quad (15)$$

The OM2 constraints are included in (9) to (14) (the same as for OM1). The results of OM2 are compared with the OM1 solutions in Figures 8 and 9. As before, the AEROVIS visualization tool is used to show the results. In OM2 the carbon emissions are not minimized, although to allow the comparison, carbon emissions are computed after the optimization run to enable these solutions to be shown in the same layout scheme. Figure 8compares the OM1 solutions (Figure 8(a)) with the OM2 solutions (Figure 8(b)) in terms of cost and carbon emissions. Two solutions are marked inside a black box. The cost of these solutions is practically the same (4.89×10^6 USD versus 4.86×10^6 USD), but the OM1 solution has 722 tonnes less CO_2 emissions than the OM2 solution. Figure 9 shows the comparisons of solutions, in terms of carbon emissions and pressure deficits, identified by OM1 (Figure 9(a)) with the OM2 (Figure 9(b)). Two solutions are also marked inside a red box. This means they have almost the same value of pressure deficits (146 m versus 145 m), but the OM1 solution has 948 tonnes less CO_2 emissions than the OM2 solution.

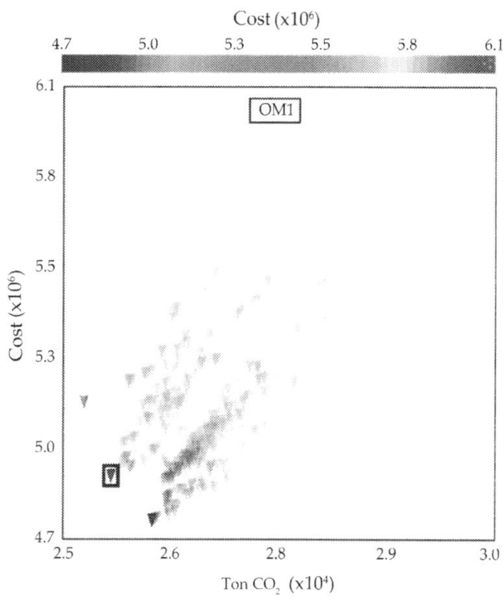

(a)

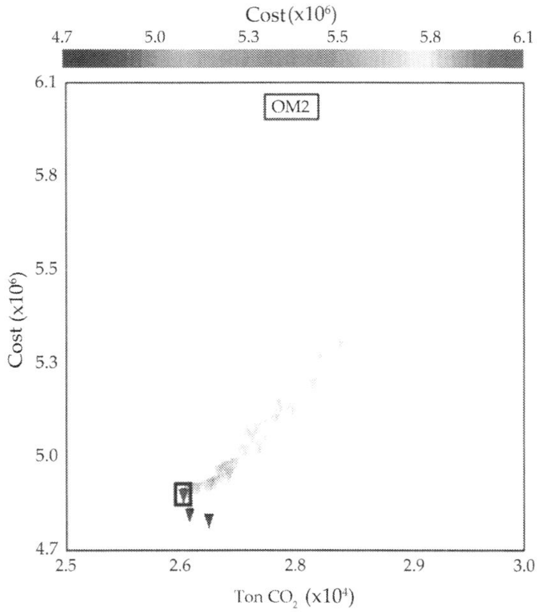

(b)

Figure 8: (a) Cost versus ton CO_2 for OM1; (b) cost versus ton CO_2 for OM2.

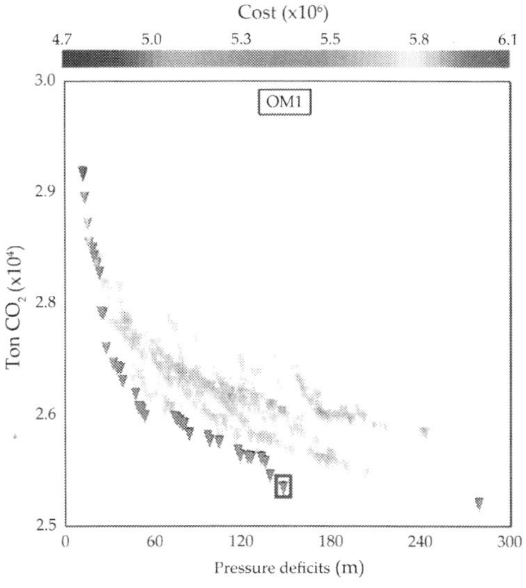

(a)

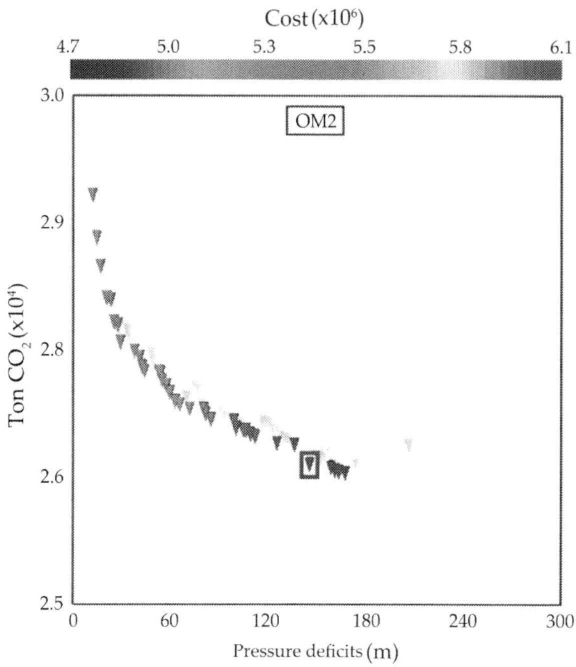

(b)

Figure 9: (a) Ton CO_2 versus pressure deficits for OM1; (b) ton CO_2 versus pressure deficits for OM2.

These results show the advantage of using a three-objective optimization model (OM1) to design WDNs. The ROs approach used here includes a multiphased design of the WDN that has to define the initial design (for the first phase) and helps decision makers to investigate different future options during the planning horizon. The results for the case study that are presented in Figure 4 show the values for the three objective functions for each solution. Each of these solutions can be detailed according to the effective WDN design for the different possible future options.

The effective design of three different WDN solutions of the Pareto front is presented in Figure 10, taking the decision path of scenario 1 into account, in which all the expansion areas are developed. Although only this scenario is represented, the approach presented does identify the designs for all the other scenarios. Furthermore, these designs are presented according to the three phased interventions in the network. The first design phase (t=1) is in fact the solution that has to start being implemented now, taking the future expansion options into account. In this phase, pipes 2 to 8 are designed with enough hydraulic capacity to cope with all possible future scenarios. Nonetheless, the case study includes a pumping station downstream of reservoir R1 (see Figure 2) that can be used to increase the network node pressures, if required. There is therefore a compromise between increasing the hydraulic capacity of pipes with an initially high construction investment or increasing the head of the pumps and spending more on energy.

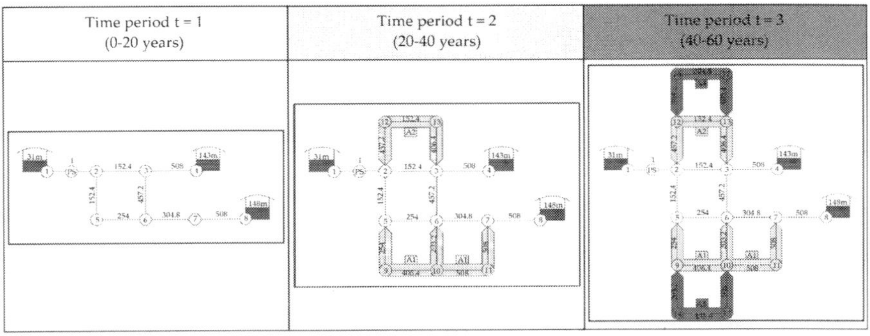

(a)

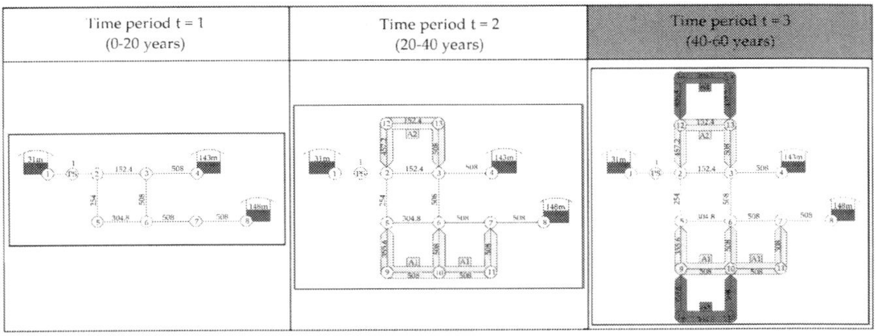

(b)

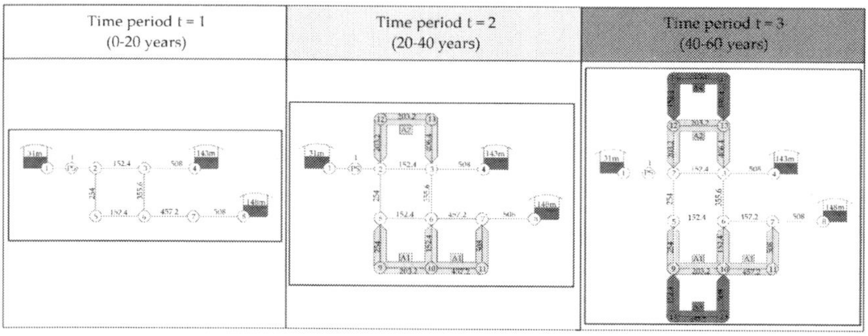

(c)

Figure 10: Water distribution network considering scenario 1 and the lowest cost solution (a); the lowest pressure deficit solution (b); and the lowest carbon emissions solution (c).

Figure 10 is divided into three to show the solutions for the phased designs, using light grey for the network areas to be installed in phase (t=2) and dark grey for the areas that will be installed in phase (t=3). Figure 10(a) shows the lowest cost design solution. Figure 10(b) shows a solution with the lowest hydraulic pressure deficit, and Figure 10(c) represents a solution with the lowest carbon emissions.

It is possible to draw some conclusions from Figure 10. In terms of pipe diameters, the solution with the lowest cost uses the smallest pipe diameters (Figure 10(a)). Conversely, the solution with large pipe diameters is also the most expensive solution and has the lowest pressure deficit (Figure 10(b)). To reduce the pressure deficit, the pipe diameters need to increase, which in turn will increase the cost and also the carbon emissions. Finally, low carbon emissions can be obtained with a low cost solution, but with high hydraulic pressure deficits (Figure 10(c)). It should be emphasized that Figure 10 shows only the design of the WDN for one of the possible scenarios (scenario 1). However, all the solutions of the Pareto front, which are obtained by the multiobjective model, include the designs for all the other scenarios.

Summing up, the results and the comparisons set out in this work comprise the optimal solutions in terms of the diameter of the pipes to be laid in each of the design phases. The proposed approach has the advantage of offering stakeholders trade-offs between the objectives of cost, pressure deficits and carbon emissions, while at the same time supporting decision making by identifying solutions for the first design phase and keeping the temporal development of WDNs in sight. Furthermore, comparison of the results of the OM1 and OM2 models shows that there are advantages in including carbon emissions as a specific objective in OM1, which can find solutions with lower carbon emissions for the same level of costs and same level of pressure deficits as the OM2 solutions.

CONCLUSIONS

This paper describes a ROs multiobjective and multiphase optimization approach for the flexible design and operation of WDNs under uncertainty. The proposed optimization model minimizes the costs, the extent of hydraulic pressure deficits and the total carbon emissions. A case study has been used to validate the approach and the results are presented in a Pareto front by using an appropriate visualization tool. The Pareto front surfaces obtained show the trade-off between the three objectives. The geometry of Pareto fronts enables some interesting conclusions to be drawn in terms of the possible compromise between the objectives and how a decrease achieved in one objective influences the increase in the others. For example, to achieve a low pressure deficit solution, the rate of increase in the required investment must grow much faster beyond certain pressure deficit threshold (60 m). On the other hand, carbon emissions do not appear to be in conflict with the cost objective as reducing the pipe material quantities and the energy consumed by pumps lowers both the cost and the carbon emissions. Finally, the results show that pressure deficit reduction can only be achieved by increasing carbon emissions.

The results of this paper also stem from comparisons between two models to ascertain the importance of including carbon emissions minimization as an objective in the optimization model. The results show that it is possible to reduce the CO_2 emitted for the same level of capital expenditure or the same level of network pressure deficits if carbon emissions are minimized in the optimization process. Finally, three particular solutions for the network design, retrieved from the edges of the Pareto front, are detailed to compare the pipe size changes when preference is given to one of the objectives. The results indicate the increase in pipe sizes observed for a solution with the lowest demand deficits relative to solutions obtained for the lowest cost and with the lowest carbon emissions. From a decision-making point of view, results of this kind can help decision makers to better understand the impact of assigning more importance to one objective relative to another.

The consideration of ROs permits uncertainty to be included in the decision process by using phased planning that makes it possible to undertake midcourse adjustments or make additional investments. It should be emphasized that the multiphase design of WDNs enables decision makers to postpone interventions in the network until such time as they are required, thus reducing the construction costs at year zero compared with the traditional single-phase design. The methodology proposed in this paper supports decisions taken for the first phase that can be adjusted if required, and as new information becomes available. The adjustments are predefined in a decision tree where different scenarios with specific occurrence probability are stated. However, given the large uncertainty in such long planning horizon, the process can be reassessed at the end of each phase in response to inaccuracies in the scenario or demand forecasts. The use of ROs gives a measure of flexibility to decision making.

ACKNOWLEDGMENTS

This study had the support of Fundação para a Ciência e Tecnologia (FCT), through the Strategic Project UID/MAR/04292/2013 granted to MARE.

REFERENCES

1. S. C. Myers, "Determinants of corporate borrowing," Journal of Financial Economics, vol. 5, no. 2, pp. 147–175, 1977. · ·
2. D. Wörsdörfer, S. Lier, and N. Crasselt, "Real options-based evaluation model for transformable plant designs in the process industry," Journal of Manufacturing Systems, vol. 42, pp. 29–43, 2017. ·
3. J. A. Schachter and P. Mancarella, "A critical review of Real Options thinking for valuing investment flexibility in Smart Grids and low carbon energy systems," Renewable and Sustainable Energy Reviews, vol. 56, pp. 261–271, 2016. · ·

4. J. Savolainen, "Real options in metal mining project valuation: review of literature," Resources Policy, vol. 50, pp. 49–65, 2016. ·
5. J. Buurman and V. Babovic, "Adaptation Pathways and Real Options Analysis: an approach to deep uncertainty in climate change adaptation policies," Policy and Society, vol. 35, no. 2, pp. 137–150, 2016. · ·
6. M. Woodward, Z. Kapelan, and B. Gouldby, "Adaptive flood risk management under climate change uncertainty using real options and optimization," Risk Analysis, vol. 34, no. 1, pp. 75–92, 2014. · ·
7. D. Huang, K. Vairavamoorthy, and S. Tsegaye, "Flexible design of urban water distribution networks," in Proceedings of the World Environmental and Water Resources Congress, pp. 4225–4236, Providence, RI, USA, May 2010.
8. I. Basupi and Z. Kapelan, "Flexible water distribution system design under future demand uncertainty," Journal of Water Resources Planning and Management, vol. 141, no. 4, Article ID 04014067, 2015. · ·
9. I. Basupi and Z. Kapelan, "Evaluating flexibility in water distribution system design under future demand uncertainty," Journal of Infrastructure Systems, vol. 21, no. 2, Article ID 4014034, 2015. · ·
10. J. Marques, M. Cunha, and D. A. Savić, "Multi-objective optimization of water distribution systems based on a real options approach," Environmental Modelling and Software, vol. 63, pp. 1–13, 2015. · ·
11. E. Alperovits and U. Shamir, "Design of optimal water distribution systems," Water Resources Research, vol. 13, no. 6, pp. 885–900, 1977. · ·
12. D. Savic, "Single-objective vs. Multiobjective optimisation for integrated decision support," in Proceedings of the 1st Biennial Meeting of the International Environmental Modelling and Software Society, pp. 7–12, 2002.
13. L. M. Herstein, Y. R. Filion, and K. R. Hall, "Evaluating the environmental impacts of water distribution systems by using EIO-LCA-based multiobjective optimization," Journal of Water Resources Planning and Management, vol. 137, no. 2, pp. 162–172,

2011. · ·

14. M. D'Ercole, M. Righetti, R. M. Ugarelli, L. Berardi, and P. Bertola, "An integrated modeling approach to optimize the management of a water distribution system: improving the sustainability while dealing with water loss, energy consumption and environmental impacts," Procedia Engineering, vol. 162, pp. 433–440, 2016. ·
15. S. Kirkpatrick, J. Gelatt, and M. P. Vecchi, "Optimization by simulated annealing," American Association for the Advancement of Science. Science, vol. 220, no. 4598, pp. 671–680, 1983. · · ·
16. N. Metropolis, A. W. Rosenbluth, M. N. Rosenbluth, A. H. Teller, and E. Teller, "Equation of state calculations by fast computing machines," The Journal of Chemical Physics, vol. 21, no. 6, pp. 1087–1092, 1953. · ·
17. J. A. Zeferino, A. P. Antunes, and M. C. Cunha, "An efficient simulated annealing algorithm for regional wastewater system planning," Computer-Aided Civil and Infrastructure Engineering, vol. 24, no. 5, pp. 359–370, 2009. · ·
18. L. M. Nunes, E. Paralta, M. C. Cunha, and L. Ribeiro, "Comparison of variance-reduction and space-filling approaches for the design of environmental monitoring networks," Computer-Aided Civil and Infrastructure Engineering, vol. 22, no. 7, pp. 489–498, 2007.
19. M. D. C. Cunha and J. Sousa, "Hydraulic infrastructures design using simulated annealing," Journal of Infrastructure Systems, vol. 7, no. 1, pp. 32–39, 2001. · ·
20. J. Reca, J. Martínez, C. Gil, and R. Baños, "Application of several meta-heuristic techniques to the optimization of real looped water distribution networks," Water Resources Management, vol. 22, no. 10, pp. 1367–1379, 2008. · ·
21. J. Reca, J. Martínez, R. Baños, and C. Gil, "Optimal design of gravity-fed looped water distribution networks considering the resilience index," Journal of Water Resources Planning and Management, vol. 134, no. 3, pp. 234–238, 2008. · ·
22. J. B. Kollat, P. M. Reed, and R. M. Maxwell, "Many-objective groundwater monitoring network design using bias-aware ensemble Kalman filtering, evolutionary optimization, and visual analytics," Water Resources Research, vol. 47, no. 2, pp. 1944–7973, 2011. · ·

23. J. Marques, M. Cunha, and D. A. Savić, "Using real options for an eco-friendly design of water distribution systems," Journal of Hydroinformatics, vol. 17, no. 1, pp. 20–35, 2015.
24. M. S. Morley and C. Tricarico, "Pressure driven demand extension for EPANET (EPANETpdd)," Tech. Rep., Centre for Water Systems, University of Exeter, Exeter, UK, 2008.
25. L. A. Rossman, "Epanet 2 users manual," Tech. Rep. EPA/600/R-00/057, U.S. Environmental Protection Agency, Washington, DC, USA, 2000.
26. E. Todini, "A more realistic approach to the 'extended period simulation' of water distribution networks," in Proceedings of the Advances in Water Supply Management Symposium (CCWI ‹03), pp. 173–184, London, UK, September 2003.
27. O. Giustolisi, D. Savic, and Z. Kapelan, "Pressure-driven demand and leakage simulation for water distribution networks," Journal of Hydraulic Engineering, vol. 134, no. 5, pp. 626–635, 2008. · ·
28. O. Giustolisi and T. M. Walski, "Demand components in water distribution network analysis," Journal of Water Resources Planning and Management, vol. 138, no. 4, pp. 356–367, 2012. · ·
29. J. M. Wagner, U. Shamir, and D. H. Marks, "Water distribution reliability: simulation methods," Journal of Water Resources Planning and Management, vol. 114, no. 3, pp. 276–294, 1988. · ·
30. T. M. Walski, J. Gessler, and J. W. Sjostrom, "Water distribution systems: simulation and sizing," in Environmental Progress, M. Wentzel, Ed., Lewis Publishers, Chelsea, UK, 1990.
31. W. Wu, A. R. Simpson, and H. R. Maier, "Accounting for greenhouse gas emissions in multiobjective genetic algorithm optimization of water distribution systems," Journal of Water Resources Planning and Management, vol. 136, no. 2, pp. 146–155, 2010. · ·
32. J. B. Kollat and P. Reed, "A framework for visually interactive decision-making and design using evolutionary multi-objective optimization (VIDEO)," Environmental Modelling and Software, vol. 22, no. 12, pp. 1691–1704, 2007.

CHAPTER 8

POWER AND EXECUTION TIME OPTIMIZATION THROUGH HARDWARE SOFTWARE PARTITIONING ALGORITHM FOR CORE BASED EMBEDDED SYSTEM

Siwar Ben Haj Hassine,[1] Mehdi Jemai,[1] and Bouraoui Ouni[1,2]

[1]Laboratory of Electronic and Microelectronic, Faculty of Sciences at Monastir, University of Monastir, 5000 Monastir, Tunisia

[2]Networked Objects Control & Communication Systems Laboratory, National Engineering School of Sousse, BP 264, Sousse Erriadh, 4023 Sousse, Tunisia

ABSTRACT

Shortening the marketing cycle of the product and accelerating its development efficiency have become a vital concern in the field of embedded system design. Therefore, hardware/software partitioning has become one of the mainstream technologies of embedded system development since it affects the overall system performance. Given today's largest requirement for great efficiency necessarily accompanied by high speed, our new algorithm presents the best version that can meet such unprecedented levels. In fact, we describe in this paper an algorithm that is based on HW/SW partitioning which aims to find the best tradeoff between power and latency of a system taking into consideration the dark silicon problem. Moreover, it has been tested and has shown its efficiency compared to other existing heuristic well-known algorithms which are Simulated Annealing, Tabu search, and Genetic algorithms.

INTRODUCTION

The exponential rise of embedded systems, all along with the persistent quest for higher levels of performance have resulted in the necessity of creating efficient types of embedded circuits. In fact, the embedded systems have become the worldwide leader technologies since they have penetrated into the human life to a very large extent. Besides, they play a vital role in industries as well as military applications which requires the necessity of having faster and better performing systems. Unfortunately, most of current technologies have only managed to further increase the system's capacity in order to have a faster treatment at the cost of a considerable simultaneous augmentation in their power. However, excessive power consumption may damage the integrated circuits through overheating, limiting the degree of transistors integration on a chip, bringing problem signal integrity, shortening battery durability for portable devices, and requiring expensive cooling and packaging systems. Moreover, the huge dependence of wastage power consumption on threshold voltage has limited further threshold and provide voltage scaling. Thus, the power consumption is rising with technology scaling, such that it can no longer be cooled down profitably considering the physical limitations forced by cooling technologies and packaging. This gives rise to the dark silicon problem [1–3]. The concept of dark silicon is based on constraint that important fraction of transistors on chip cannot be powered on at a nominal voltage for a specific thermal design power TDP budget and have to be power-gated or simply remain dark. The TDP is the maximum amount of power provided to a chip while maintaining the chip temperature under the thermal safe temperature. In case where the TDP is exceeded, the temperature of the chip will rise beyond the cooling capacity that will throttle the chip. Previous studies [1, 2] have predicted that 50% to 80% of the chip area will be dark for GPU and CPU based systems. To overcome such dilemmas, designers' efforts have been increased to produce less consuming systems. In this context, some research groups have focused on the creation of new architectures in terms of the material [4] while other groups have focused on extending batteries life cycle [5]. Yet, such solutions require high resources that several research groups do not have. For that, other methods have appeared in order to offer a less power consuming system such as hardware/software partitioning [6, 7].

Traditionally, partitioning was carried out manually which requires a detailed knowledge of circuit operations from designers. Such manual approaches were limited only to small designs with small number of constituent blocks [8, 9]. Since digital systems have become much more sophisticated, automatic HW/SW partitioning has become a necessity. In fact, many research groups have opted for the HW/SW partitioning in order to increase the performance of a system as presented in approaches such as [10, 11]; most of these approaches purposes are to meet performance constraints while keeping the system cost (area) as low as possible. Unfortunately, none of them took the power consumption and the execution time into consideration. Hence, we present in this paper an algorithm that finds a possible HW/SW partitioning of a data flow graph that finds out a tradeoff between power and latency taking into account the dark silicon problem.

The rest of the paper is organized as follows: Section 2 reviews the related literature; the proposed partitioning algorithm is addressed in Section 3 followed by an illustrative example; the numerical experimentation and discussion theorem are presented in Section 5 and finally the article ends up with the conclusion that briefs the present findings and future research on this theme.

RELATED WORK

Recently, a new alternative technology that combines logic elements and memory along with an intellectual property processor core has emerged to remedy the excessive need for better performance systems. This technology called System on Programmable Chip SoPC allows and facilitates the SW/HW partitioning.

As generally reckoned, embedded systems consist of programmable software part (SW) and application specific hardware part (HW). Software part is much easier to develop and modify, and it consumes less power compared to the hardware part but it requires extra time to give final response. In fact, compared to the software which is less expensive in terms of cost and power consumption, the hardware provides better performance because it offers a faster treatment. For that reason, the purpose of HW/SW partitioning is to design a balanced system that accomplishes all system constraints [12]. Most of formulations of HW/SW partitioning dilemma have proven to be NP-hard [13, 14]. In fact, many exact algorithms have been proposed such as Branch-and-Bound [15], dynamic programming [16], and integer linear programming [17]. However, these exact algorithms tend to be quite slow for bigger inputs. Hence, for bigger partitioning problem heuristic algorithms have been the basis for the majority of researches such as Genetic algorithm (GA) [18], Tabu Search [19, 20], Simulated Annealing [21], Particle Swam Optimization [22, 23], Ant algorithm [24, 25], shuffled frog leaping algorithm [26], and greedy algorithm [27]. Other designers have mixed two heuristic algorithms to solve the HW/SW partitioning problems like in [28] where authors have used hybrid algorithm of Genetic algorithm (GA) and Tabu Search one, while others [29] have mixed the Discrete Particle Swarm Optimization (DPSO) and Branch-and-Bound (B&B) algorithms to meet the same aim. Besides, authors in [30] have proposed a new heuristic solution based on HW/SW partitioning that aims to reduce the execution time of the overall circuit. Moreover, authors in [31] have come up with a new IVA-HD which is a programmable, true multistandard, and full HD video coding engine that adopts HW/SW partitioning to achieve the low power and area equipment of the OMAP 4 processor. To attain the same goal of power optimization, [32] has proposed a minimizing approach based on mapping clusters of instructions to a core that yields a high utilization rate of resources and thus minimizes power consumption. Such a method has offered a less consuming system at the cost of an additional hardware overhead. The problems that these previously mentioned works have met are either to optimize one parameter at the cost of another important constraint or to focus on achieving the optimization of only one constraint such as power or execution time. Also, none of them have mentioned the dark silicon problem. In fact, the dark silicon has become a critical issue for designers since it can decrease the reliability in the nanoera [33–35] and leads to soft errors, aging and even process variations [36, 37]. Recent works have explored the dark silicon problem by applying a very low voltage to power on more cores [38] and proposed new accelerators architectures [39, 40]. Almost, the majority of works have handled the dark silicon problem on low level codesign which necessitates a good knowledge of the target circuit and extra time of marketing cycle of the product. Other designers have proposed new architectures by exploiting architectural heterogeneity [41–43]. However, such solutions require high resources that several research groups do not have. In the literature, only few works have combined the HLS and dark silicon problem due to its complexity [44, 45]. It is true that generally the dark silicon problem appears for multiprocessor system-on-chip (MPSoc). But, due to the Soc huge rise the dark silicon problem must be taken into consideration even with one core based embedded system [46]. Motivated by the fact and coming across the shortages of other researches, it has been vital to come up with a new idea of developing a new algorithm that aims to create a less consuming system and a faster one without influencing the system reliability.

PROBLEM'S DEFINITIONS

We consider the applications that can be modeled using data flow graph (DFG). A data flow graph that is used to create a preliminary overview of the system denoted as (V, E), where $V=\{V_1, V_i, \ldots, V_n\}$, is the set of vertices or nodes that are interconnected to each other by edges $E = \{\{V_i, V_j\}, \{V_i, V_{j+1}\}, \ldots\}$. Edges of the graph present the dependencies between the components of the system. In general, the node of the graph can represent a basic block [47], a short of instruction [48], a procedure or a function [49], and so on. In this paper, we use four different types of nodes:

- A start or an end node $V_i^{startend}$; $V_i^{startend} \in V^{startend}$ and $V^{startend} \subseteq V$.
- A node that includes simple code V_i^S; $V_i^S \in V^S$ and $V^S \subseteq V$.
- A node that contains the beginning of a control construct V_i^B; $V_i^B \in V^B$ and $V^B \subseteq V$.

- A node that contains the end of a control-construct V_i ; $V_i^E \in V^E$ and $V^E \subseteq V$.

Partitions' Types

The graph partitioning is to cut the graph into possible partitions $P_{all} = \{P_1, P_i,...,_n\}$ where P_{all} is the set of all possible partitions; P_i is a possible partition; and n is the number of possible partitions.

There exist two kinds of partitions:

- A control-construct partition that includes a whole construct such as if to end if, case to end case, and so on.
- A mix partition that could contain either two or more control-constructs or one or more control-construct combined with a simple node (that contains simple construct such as addition operation).

Node's Links

To facilitate the search of control-construct partitions, we have used the parameter of link. If the node is a beginning of a control-construct or an end of a control construct, the link value equals 1. For the rest of node's types, the link value equals 0. The link definition can be defined as follows:

$$\text{Link}\left(v_i\right)\begin{cases} = 1, & \text{if } v_i \in V^B \text{ or } v_i \in V^E \\ 0, & \text{else.} \end{cases} \tag{1}$$

Related Statements

When a task is realized by hardware or software, its execution time and power consumption show diverse values. We define the following functions $H_L(P_i)$, $S_L(P_i)$, $H_P(P_i)$ and $S_P(P_i)$ to represent the hardware latency, the software latency, the hardware power, and the software power respectively of a given partition P_i. Although obtaining the exact values of the execution time and power consumption is a challenging problem, it is beyond the scope of this article. Rather, we focus on algorithmic issues in partitioning.

Given a path P_w: $V_i \rightarrow V_j$ and a hardware/software partitioning P_i for all the nodes in P_w, the completion time of P_w under partitioning P_i is the summation of all the latencies occurred on P_w taking into consideration the parallel execution of some tasks. The system completion time is defined to be the completion time of a critical path Cp in DFG. The hardware latency and the software latency corresponding to a target partition can be written as follows:

$$H_L(P_i) = H_L(C_P(P_i)) = \max_{v_i \in P_i} \left[(X_m(i) H_L(v_i)) + \sum_{v_j \in P_i} ((X_m(j) * \psi_{ij}) H_L(v_j)) \right], \tag{2}$$

where $\{\psi_{ij} = 1$, if V_j depends to V_i; 0, else$\}$ and $\{X_m(i) = 0$, if V_i is executed by Software; $X_m(i) = 1$, if V_i is executed by Hardware$\}$;

$$S_L(P_i) = \sum_{v_i \in P_i} S_L(v_i). \tag{3}$$

For a given (V, E), we define a vector X_m to indicate either the task is realized by hardware or software. For instance, for a node V_i, $X(i)$ equals 0 if the task is executed by the software and equal to 1 if it is realized by the hardware.

The power consumption of the system with respect to a given partitioning can be calculated as the summation of all the task power consumption of each node realized by software or hardware. In fact, it can be written as follows:

$$H_P(P_i) = \sum_{v_i \in P_i} H_P(v_i),$$

$$S_P(P_i) = \sum_{v_i \in P_i} S_P(v_i). \tag{4}$$

So, to recapitalize, we define the hardware/software partitioning problem as follows: given (V, E) and thermal design power, find a partitioning that offers the best tradeoff between total power and execution time of the system.

PROPOSED ALGORITHM

Our algorithm is meant to achieve graph partitioning in order to find the best compromise between power and execution time. As generally reckoned, the software consumes less power than the hardware but it requires more time to give response while the hardware which tackles the problem of timing consumes more power. This approach starts with a system totally implemented by software, it will not consume power but it will be too tardy. Whenever a partition of the system migrates to be executed by the hardware, the system will consume more power and become faster. As mentioned previously, our algorithm includes two different kinds of partitions. Its first function is to search for all control-construct partitions (Algorithm 2) and then it builds the mix partitions. After that, it makes all possible combinations between the generated partitions (Algorithm 1).

// generate the mix partitions
For all $P_i \in P_{all}$
    If $\forall_{v_i \in V}$ Link(pred $(v_{startp}) = 0)$ **then** // v_{startp} is the start node of the control-construct partition
        $P_k \leq \{\text{pred}(v_{startp}), P_i\}$
            If $(P_k \nexists)$
                $P_{all} \leq P_k$
                $k = k + 1;$
            End if

    else
            If $(P_{i-1} \in P_{control})$
                $P_k \leq \{P_{i-1}, P_i\};$
                If $(P_k \nexists)$
                    $P_{all} \leq P_k$
                    $k = k + 1$
                End if
            End if
    End if
    If $\forall_{v_i \in V}$ Link(succ$(v_{endp}) = 0)$ **then** // v_{endp} is the end node of the control-construct partition
        $P_k \leq \{Pi; \text{succ}(v_{endp})\}$
            If $(P_k \nexists)$
                $P_{all} \leq P_k$
                $k = k + 1;$

```
        else
                If (P_{i+1} ∈ P_control)
                  P_k ≤ {P_i, P_{i+1}};
                  If (P_k ∄)
                    P_all ≤ P_k
                    k = k + 1
                  End if
                End if
      End if
    End for;
    // find the combination of the previous generated partitions
    P_test ≤ P_all
    For all P_i ∈ P_test
        z ≤ i + 1
        For all P_z ∈ P_test

          if P_i ⊈ P_z then
            P_k ≤ {P_i, P_z};
            P_all ≤ P_k
            z ≤ z + 1
            k ≤ k + 1
          else
            z ≤ z + 1
          end if
    end for
    if P_control = {Ø} then
      while (P_i ≠ V) do
        P_k ≤ make all possible combinations
        P_all ≤ P_k

        k = k + 1
      end
    end if
    End begin
```

Algorithm 1: Generation of partitions.

Begin
For all $v_i \in V$
Compute Link(v_i)
End for
$k = 1;\ k \in \mathbb{N}$
// generation of all control construct partitions
For all $v_i \in V^B$:
If $\forall_{v_i, v_j \in V}$ (Link(v_i) = 1 and $v_j \notin V^B$) then
$P_k \leq v_i$
do
$P_k \leq v_j$;
While (Link(v_j) = 1)
$P_{\text{all}} \leq P_k$
$P_{\text{control}} \leq P_k$
$k = k + 1$
End if
End for

Algorithm 2: Generation of control-construct partitions.

In case where all nodes are simple, the partitions will simply take all possible combinations of the nodes. Our algorithm (Algorithm 3) is based on three functions. F_1 provides the total latency of the system for each generated partition while F_2 computes the total consumed power of the system under a given partition. When the algorithm becomes so close to the best solution, it will offer us an interval that includes some suggestions. F_1 and F_2 are written as follows:

$$F_1\,(Pi) = \left[S_L\,(C_P) - S_L\,(C_P\,(Pi)) + H_L\,(C_P\,(Pi))\right], \tag{5}$$

where $S_L(C_P) = \sum_{\text{Vi}\in Cp} S_L(\text{V}_i)$, $S_L(C_P(Pi))$ and $H_L(C_P(Pi))$ are the software and the hardware latency of the critical path of the partition P_i, respectively.

$$F_2\,(P_i) = \left[S_P\,(G) - S_P\,(Pi) + H_P\,(Pi)\right], \tag{6}$$

where $S_L(\text{G}) = \sum 6_{\text{Vi}\in Cp}\, S_{\text{p}}(\text{V}_i)$

Begin
Find all paths P_w
Calculate $S_L(\text{all}) \leq \{S_L(P_{w1}), S_L(P_{w2}), \ldots, S_L(P_{wn})\}$
Find Cp
Compute $S_L(C_P)$
Generate all partitions
For all $\mathbf{P}_i \in P_{\text{all}}$ **do**
Find $C_P\,(Pi)$
if $(v_i \in P_i) \not\subseteq$ Cp **then**
$F_1 = S_L(C_P)$
else

$S_L(\text{new}) \leq S_L(\text{all})$ // $S_L(\text{new})$ is the set of the new latency of each path
For all $\mathbf{P}_w$ **do**
Compute $S_L(P_w)$; // the new latency of the path $S_L(P_w) = F_1(P_w)$;
end for
update $S_L(\text{new})$
$F_1(Pi) \leq \max[S_L(\text{new})]$
end if
Compute function $F_2(P_i)$
End for
// choose the best partitions//

$m \leq 1$
TDP $\leq$ the dark power
For all $\mathbf{P}_i \in P_{\text{all}}$ **do**
For all $v_h \in P_i$ **do**
if ($v_h \in$ Cp and $v_{h+1} \notin$ Cp) **then**
if $F_2(P_i) <$ TDP **then**
$P_{\text{suggested}} \leq P_i$
end if
end for
For all $\mathbf{P}_i \in P_{\text{suggested}}$ **do**
Compute function $F_3\,(P_i)$

End for

$m \leq$ number of partition

For i from 1 to m **do**

$$P_{best} \leq \frac{1}{m} \times \sum_{i=1}^{m} F_3\,(Pi)$$

End for

End.

Algorithm 3: Our proposed algorithm.

To avoid the dark silicon problem, we have introduced a new constraint called thermal design power TDP. This constraint refers to the maximum amount of power that can be provided to a chip while maintaining the chip temperature under the thermal safe temperature. If the system power under a specific partition exceeds the TDP, then that partition will not be among the suggested ones. Thus, ensure the good performance and the reliability of the system.

F_3 is introduced to facilitate taking the decision of which the suggested partition is the best one. F_3 is introduced as follows:

$$F_3\,(Pi) = \frac{F_2\,(Pi)}{F_1\,(Pi)}. \quad (7)$$

The best solution equals the closest value to

$$\frac{1}{m} \times \sum_{i=1}^{m} F_3\,(Pi), \quad (8)$$

where m is the number of suggested solutions.

ILLUSTRATIVE EXAMPLE

To further clarify our algorithm, it has been applied on the graph shown in Figure 1. The node's constraints $V = \{V_1, V_i, \ldots, V_{15}\}$ are presented in Table 1.

Table 1: Node's parameters

	v_1	v_2	v_3	v_4	v_5	v_6	v_7	v_8	v_9	v_{10}	v_{11}	v_{12}	v_{13}	v_{14}	v_{15}
H_L	3	2	1	2	1	2	3	1	1	3	3	1	5	4	2
S_L	6	4	2	4	2	4	7	2	2	7	6	2	10	9	5
H_P	2	1	2	1	1	2	2	1	1	3	3	1	4	4	5
S_P	0.5	0.25	0.5	0.25	0.25	0.5	0.5	0.25	0.25	1	1	0.25	1	1.5	2

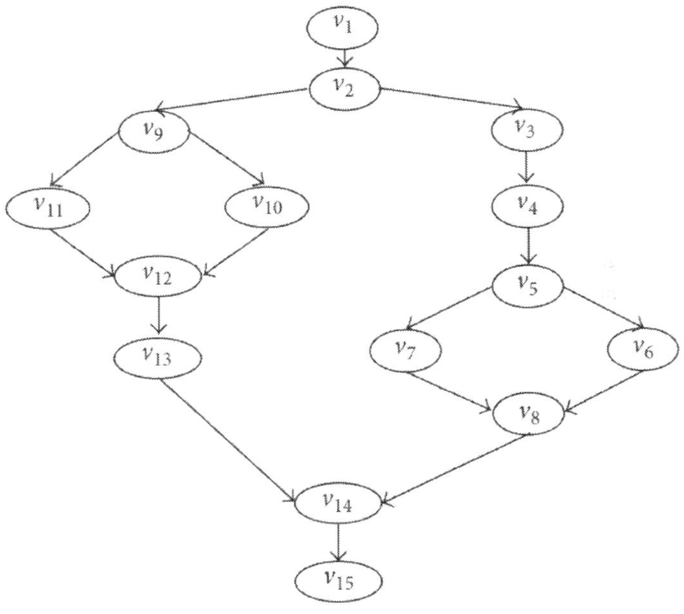

Figure 1: The data flow graph.

The first step consists in finding all possible paths and calculating the software latency of all paths in order to get the critical path. In our case C_p: $S(C_p) = 45$. The second step is to compute the link of each node as Table 2 presents. Then, all possible partitions will be generated.

Table 2: The links of nodes

	v_1	v_2	v_3	v_4	v_5	v_6	v_7	v_8	v_9	v_{10}	v_{11}	v_{12}	v_{13}	v_{14}	v_{15}
Link	0	1	0	0	1	0	0	1	1	0	0	1	0	1	0

In our case, there exist eleven partitions $P_{all} = \{P_1, P_2, ...,_{11}\}$ as Table 3 describes.

Table 3: The generated partitions

Partitions	Nodes														
P_1	v_1	v_2	v_3	v_4	$\mathbf{v}_5$	$\mathbf{v}_6$	$\mathbf{v}_7$	$\mathbf{v}_8$	v_9	v_{10}	v_{11}	v_{12}	v_{13}	v_{14}	v_{15}
P_2	v_1	v_2	v_3	v_4	v_5	v_6	v_7	v_8	$\mathbf{v}_9$	$\mathbf{v}_{10}$	$\mathbf{v}_{11}$	$\mathbf{v}_{12}$	v_{13}	v_{14}	v_{15}
P_3	v_1	v_2	v_3	$\mathbf{v}_4$	$\mathbf{v}_5$	$\mathbf{v}_6$	$\mathbf{v}_7$	$\mathbf{v}_8$	v_9	v_{10}	v_{11}	v_{12}	v_{13}	v_{14}	v_{15}
P_4	v_1	v_2	v_3	v_4	v_5	v_6	v_7	v_8	$\mathbf{v}_9$	$\mathbf{v}_{10}$	$\mathbf{v}_{11}$	$\mathbf{v}_{12}$	$\mathbf{v}_{13}$	v_{14}	v_{15}
P_5	v_1	v_2	$\mathbf{v}_3$	$\mathbf{v}_4$	$\mathbf{v}_5$	$\mathbf{v}_6$	$\mathbf{v}_7$	$\mathbf{v}_8$	v_9	v_{10}	v_{11}	v_{12}	v_{13}	v_{14}	v_{15}
P_6	v_1	v_2	v_3	v_4	$\mathbf{v}_5$	$\mathbf{v}_6$	$\mathbf{v}_7$	$\mathbf{v}_8$	$\mathbf{v}_9$	$\mathbf{v}_{10}$	$\mathbf{v}_{11}$	$\mathbf{v}_{12}$	v_{13}	v_{14}	v_{15}
P_7	v_1	v_2	v_3	v_4	$\mathbf{v}_5$	$\mathbf{v}_6$	$\mathbf{v}_7$	$\mathbf{v}_8$	$\mathbf{v}_9$	$\mathbf{v}_{10}$	$\mathbf{v}_{11}$	$\mathbf{v}_{12}$	$\mathbf{v}_{13}$	v_{14}	v_{15}
P_8	v_1	v_2	v_3	$\mathbf{v}_4$	$\mathbf{v}_5$	$\mathbf{v}_6$	$\mathbf{v}_7$	$\mathbf{v}_8$	$\mathbf{v}_9$	$\mathbf{v}_{10}$	$\mathbf{v}_{11}$	$\mathbf{v}_{12}$	v_{13}	v_{14}	v_{15}
P_9	v_1	v_2	$\mathbf{v}_3$	$\mathbf{v}_4$	$\mathbf{v}_5$	$\mathbf{v}_6$	$\mathbf{v}_7$	$\mathbf{v}_8$	$\mathbf{v}_9$	$\mathbf{v}_{10}$	$\mathbf{v}_{11}$	$\mathbf{v}_{12}$	v_{13}	v_{14}	v_{15}
P_{10}	v_1	v_2	v_3	$\mathbf{v}_4$	$\mathbf{v}_5$	$\mathbf{v}_6$	$\mathbf{v}_7$	$\mathbf{v}_8$	$\mathbf{v}_9$	$\mathbf{v}_{10}$	$\mathbf{v}_{11}$	$\mathbf{v}_{12}$	$\mathbf{v}_{13}$	v_{14}	v_{15}
P_{11}	v_1	v_2	$\mathbf{v}_3$	$\mathbf{v}_4$	$\mathbf{v}_5$	$\mathbf{v}_6$	$\mathbf{v}_7$	$\mathbf{v}_8$	$\mathbf{v}_9$	$\mathbf{v}_{10}$	$\mathbf{v}_{11}$	$\mathbf{v}_{12}$	$\mathbf{v}_{13}$	v_{14}	v_{15}

For each generated partition the latency as well as the consumed power will be calculated using the functions F_1 and F_2 respectively (Table 4). For instance, P_1 does not belong to C_p so the latency of the system will stay the same "45." However, with partitions whose nodes belong to C_p the latency of the graph can change such as in the case of P_2 and in such circumstances we have to be aware that the critical path changes as well. The final decision is taken using $F3$. The pace of function F_1 and F_2 and the best partition are shown in Figure 2.

Table 4: The calculation results

	P_1	P_2	P_3	P_4	P_5	P_6	P_7	P_8	P_9	P_{10}	P_{11}
F_1	45	41	45	41	45	39	35	39	39	34	34
F_2	14.5	15.5	15.25	18.5	16.75	20	23	20.75	22.25	23.75	25.25
F_3	0.32	0.38	0.39	0.45	0.37	0.51	0.66	0.53	0.57	0.7	0.74
$P_{suggested}$	P_1	P_2	P_3	P_4	P_5	$\mathbf{P}_6$	$\mathbf{P}_7$	$\mathbf{P}_8$	$\mathbf{P}_9$	$\mathbf{P}_{10}$	$\mathbf{P}_{11}$
$(1/m) \times \sum_{i=1}^{m} F_3(Pi)$						0.62					
Best partition						P_7					

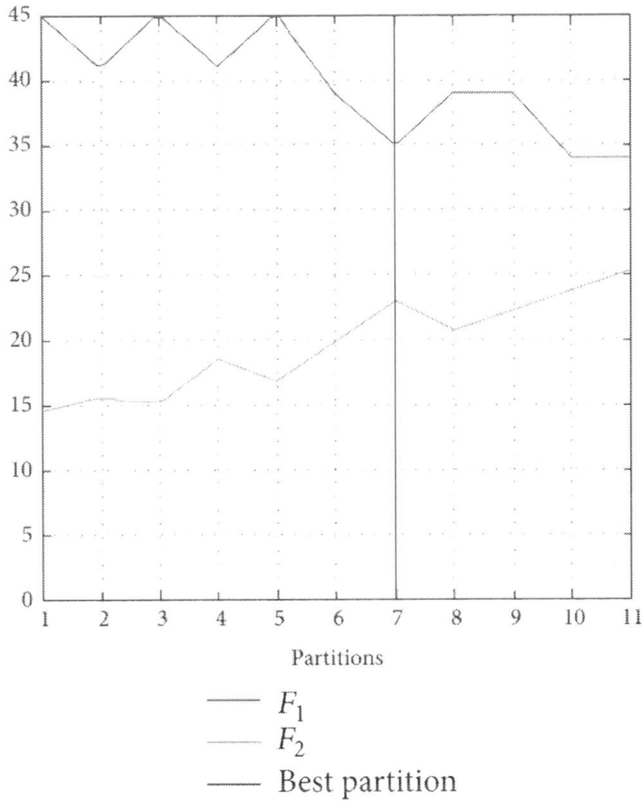

Figure 2: The pace of functions F_1 and F_2.

We have to mention that the used TDP value in the previous example is 26. In case where TDP value equals only 22, then the suggested partitions will only be the set of $\{P_6, P_8\}$ and the best partition will be P_6 instead of P_7.

EXPERIMENT RESULTS

To prove the efficiency of our algorithm, we have implemented a comparative study that is meant to illustrate the amount of power consumed by a given application and its execution time by the use of our algorithm in comparison to three existing heuristic algorithms which are the Simulated Annealing (SA), Tabu Search (TS), and Genetic algorithms (GA). To put that into practice we have applied our approach on 8-point Discrete Cosine Transform (DCT) [50] (Figure 6), 16-point DCT [51] (Figure 7), and H.264 [52]. The DCT is the most intensive part of the CLD algorithm. The H.264 is a video coding format that is one of the most used formats for compressing, recording, and distributing of video content. The characterizations of the three previous applications are presented in Table 5.

Table 5: The characterization of each application

	Number of nodes	
8-point DCT	42	
	16 multiplication operations	26 addition operations
16-point DCT	224	
	128 multiplication operations	96 addition operations
H.264	28	

The hardware power values are based on the results provided in [53] and [54] for the DCT and the H.264, respectively. The software power is almost negligible compared to the hardware power. The latency of hardware (FPGA) equals one-third to one-fifth of the latency of software (processor) [55]. The TDP value equals 7W for the rest of the comparison.

Table 6 illustrates the design results provided by our approach for the 8-point DCT, 16-point DCT, and H.264 applications.

Table 6: Design result provided by our algorithm

8-point DCT	Number of partitions		458	
	Best partition	Hardware	12 operations of (×)	19 operations of (+)
		Software	4 operations of (×)	7 operations of (+)
	Total power consumed (mW)		82.35	
	The system's latency (ns)		453	
16-point DCT	Number of partitions		12512	
	Best partition	Hardware	68 operations of (×)	91 operations of (+)
		Software	60 operations of (×)	5 operations of (+)
	Total power consumed (mW)		525.876	
	The system's latency (ns)		2780	

H.264	Number of partitions	24
	Total power consumed (W)	6.850
	The system's latency (ns)	1530

Comparison with Other Algorithms and Discussion

Tables 7, 8 and 9 summarize the design results provided by Simulated Annealing, Tabu Search, and Genetic algorithms, respectively.

Table 7: Design result provided by Simulated Annealing algorithm

8-point DCT	Best partition	Hardware	14 operations of (×)	4 operations of (+)
		Software	2 operations of (×)	22 operations of (+)
	Total power consumed (mW)		91.55	
	The system's latency (ns)		586	
16-point DCT	Best partition	Hardware	60 operations of (×)	48 operations of (+)
		Software	68 operations of (×)	48 operations of (+)
	Total power consumed (mW)		487.78	
	The system's latency (ns)		3024	
H.264	Total power consumed (W)		6.250	
	The system's latency (ns)		1740	

Table 8: Design result provided by Tabu Search algorithm

8-point DCT	Best partition	Hardware	13 operations of (×)	0 operations of (+)
		Software	3 operations of (×)	26 operations of (+)
	Total power consumed (mW)		86.8	
	The system's latency (ns)		493	
16-point DCT	Best partition	Hardware	40 operations of (×)	44 operations of (+)
		Software	88 operations of (×)	52 operations of (+)
	Total power consumed (mW)		393.712	
	The system's latency (ns)		3776	
H.264	Total power consumed (W)		5.276	
	The system's latency (ns)		1990	

Table 9: Design result provided by Genetic algorithm

8-point DCT	Best partition	Hardware	12 operations of (×)	22 operations of (+)
		Software	4 operations of (×)	4 operations of (+)
	Total power consumed (mW)		84.4	
	The system's latency (ns)		630	
16-point DCT	Best partition	Hardware	60 operations of (×)	90 operations of (+)
		Software	68 operations of (×)	6 operations of (+)
	Total power consumed (mW)		501.652	
	The system's latency (ns)		2896	
H.264	Total power consumed (W)		6.050	
	The system's latency (ns)		1650	

To compare our algorithm to the previously mentioned algorithms, we have introduced a comparison metric called Ω. It is obvious that the worst case W_{worst} presents when the system consumes the highest value of power (purely hardware) and takes too much time (purely software). We assume that $W_{worst} = H_{pmax} \times S_{Lmax}$. The best case W_{best} takes place when the system consumes less power and responds faster: $W_{best} = H_{pmin} \times S_{Lmin}$.

$$\Omega = F_1 \times F_2 = H_{psystem} \times S_{Lsystem}, \quad (9)$$

where $W_{worst} < \Omega < W_{best}$.

According to (9), when Ω is close to W_{best}, the solution is better. Table 10 presents the different values of Ω.

Table 10: Design results

	W_{worst}	**Simulated Annealing**	**Tabu Search**	**Genetic algorithm**	**Our algorithm**	W_{best}
8-point DCT	73.482	53.648	54.648	40.206	37.304	8.379
16-point DCT	3335	1475.04	1486.65	1558	1461.9	531.831
H.264	27462	10875	10499.24	10560	10480.5	3432.75

Based on Table 10, we deduce that our algorithm offers less and closest rates to the best case, whereas using Simulated Annealing, Tabu Search, and Genetic algorithms the values of Ω are too high and far from the best state. Thus, we admit that our algorithm offers the best tradeoff between the two critical parameters: power consumption and system execution time.

CONCLUSION

Given today's requirements for less consuming systems accompanied by high speed, the necessity of creating more efficient types of embedded systems has been persisted. One of the most elegant solutions that provides system's optimizations is the HW/SW partitioning. For that, we have developed a new algorithm based on HW/SW partitioning in order to obtain the best tradeoff between power and latency taking into account the dark silicon problem. Our algorithm has been applied and tested to Simulated Annealing, Tabu Search, and Genetic algorithms and as the research has illustrated, we admit that our algorithm is best suited for the urgent achievement of the desired combination of high speed and less power in core based embedded systems.

REFERENCES

1. Z. Hadi Esmaeil, E. Blem, R. A. Amant, K. Sankaralingam, and D. Burger, "Dark silicon and the end of multicore scaling," in Proceedings of the 38th Annual International Symposium on Computer Architecture (ISCA '11), pp. 365–376, ACM, San Jose, Calif, USA, June 2011.
2. N. Hardavellas, M. Ferdman, B. Falsafi, and A. Ailamaki, "Toward dark silicon in servers," IEEE Micro, vol. 31, no. 4, pp. 6–15, 2011.
3. M. B. Taylor, "Is dark silicon useful?: harnessing the four horsemen of the coming dark silicon apocalypse," in Proceedings of the 49th Annual Design Automation Conference (DAC ‹12), pp. 1131–1136, ACM, San Francisco, Calif, USA, June 2012.
4. Silicon Laboratories, http://www.silabs.com/Pages/default.aspx.
5. Lawrence Berkeley National Laboratory (Berkeley Lab), http://www.lbl.gov/.

6. T. Yen and W. Wolf, Hardware-Software Co-Synthesis of Distributed Embedded Systems, Springer US, Boston, Mass, USA, 1996.
7. J. Staunstrup and W. Wo, Hardware/Software Co-Design: Principles and Practice, October 1997, https://books.google.tn/books?hl=fr&lr=&id=yKXzBwAAQBAJ&oi=fnd&pg=PR16&dq=Hardware/Software+Co-Design:+Principles+and+Practice&ots=FFzq9utmwi&sig=TiLf_EKndx8SU_Ffg6Hx32teqBw&redir_esc=y#v=onepage&q=Hardware%2FSoftware%20Co-Design%3A%20Principles%20and%20Practice&f=false.
8. F. Cloute, J.-N. Contensou, D. Esteve, P. Pampagnin, P. Pons, and Y. Favard, "Hardware/software co-design of an avionics communication protocol interface system: an industrial case study," in Proceedings of the 7th International Conference on Hardware/Software Codesign (CODES ‹99), pp. 48–52, Rome, Italy, May 1999.
9. B. Mei, P. Schaumont, and S. Vernalde, "A hardware/software partitioning and scheduling algorithm for dynamically reconfigurable embedded systems," in Proceedings of the 11th IEEE Program for Research on Integrated Systems and Circuits, Veldhoven, The Netherlands, 2000.
10. S. Dimassi, M. Jemai, B. Ouni, and A. Mtibaa, "Hardware-software partitioning algorithm based on binary search trees and genetic algorithm to optimize logic area for SOPC," Journal of Theoretical & Applied Information Technology, vol. 66, no. 3, pp. 788–794, 2014.
11. M. Jemai, S. Dimassi, B. Ouni, and A. Mtibaa, "Optimization of logic area for System on Programmable Chip based on hardwaresoftware partitioning," in Proceedings of the International Conference on Embedded Systems and Applications (ICESA ‹14), Hammamet, Tunisia, March 2014.
12. J. Teich, "Hardware/software codesign: the past, the present, and predicting the future," Proceedings of the IEEE, vol. 100, pp. 1411–1430, 2012.
13. P. Arató, Z. Á. Mann, and A. Orbán, "Algorithmic aspects of hardware/software partitioning," ACM Transactions on Design Automation of Electronic Systems, vol. 10, no. 1, pp. 136–156, 2005.

14. W. Jigang, T. Srikanthan, and G. Chen, "Algorithmic aspects of hardware/software partitioning: 1D search algorithms," Institute of Electrical and Electronics Engineers. Transactions on Computers, vol. 59, no. 4, pp. 532–544, 2009.
15. K. S. Chatha and R. Vemuri, "Hardware-software partitioning and pipelined scheduling of transformative applications," IEEE Transactions on Very Large Scale Integration (VLSI) Systems, vol. 10, no. 3, pp. 193–208, 2002.
16. J. Wu and T. Srikanthan, "Low-complex dynamic programming algorithm for hardware/software partitioning," Information Processing Letters, vol. 98, no. 2, pp. 41–46, 2006.
17. S. Banerjee, E. Bozorgzadeh, and N. D. Dutt, "Integrating physical constraints in HW-SW partitioning for architectures with partial dynamic reconfiguration," IEEE Transactions on Very Large Scale Integration (VLSI) Systems, vol. 14, no. 11, pp. 1189–1202, 2006.
18. K. Anil and C. H. Shampa, "Design optimization using genetic algorithm and cuckoo search," in Proceedings of the IEEE International Conference on Electro/Information Technology (EIT ‹11), pp. 1–5, IEEE, 2011.
19. J. Wu, P. Wang, S.-K. Lam, and T. Srikanthan, "Efficient heuristic and tabu search for hardware/software partitioning," The Journal of Supercomputing, vol. 66, no. 1, pp. 118–134, 2013.
20. J. Wu, T. Srikanthan, and T. Lei, "Efficient heuristic algorithms for path-based hardware/software partitioning," Mathematical and Computer Modelling, vol. 51, no. 7-8, pp. 974–984, 2010.
21. Y. Jing, J. Kuang, J. Du, and B. Hu, "Application of improved simulated annealing optimization algorithms in hardware/software partitioning of the reconfigurable system-on-chip," in Parallel Computational Fluid Dynamics: 25th International Conference, ParCFD 2013, Changsha, China, May 20–24, 2013. Revised Selected Papers, vol. 405 of Communications in Computer and Information Science, pp. 532–540, Springer, Berlin, Germany, 2014.
22. S.-A. Li, C.-C. Hsu, C.-C. Wong, and C.-J. Yu, "Hardware/software co-design for particle swarm optimization algorithm," Information Sciences, vol. 181, no. 20, pp. 4582–4596, 2011.·
23. J. Wu, T. Srikanthan, and G. Chen, "Algorithmic aspects of hardware/

software partitioning: 1D search algorithms," IEEE Transactions on Computers, vol. 59, no. 4, pp. 532–544, 2010.

24. T. He and Y. Guo, "Power consumption optimization and delay based on ant colony algorithm in network-on-chip," Engineering Review, vol. 33, no. 3, pp. 219–225, 2013.
25. Y.-D. Zhang, L.-N. Wu, G. Wei, H.-Q. Wu, and Y.-L. Guo, "Hardware/software partition using adaptive ant colony algorithm," Control and Decision, vol. 24, no. 9, pp. 1385–1389, 2009. ·
26. T. Zhang, X. Zhao, Y.-K. Yu et al., "Reserch on hardware/software partitioning method of improved shuffled frog leaping algorithm," Journal of Signal Processing, vol. 9, article 003, 2015.
27. G. Lin, "An iterative greedy algorithm for hardware/software partitioning," in Proceedings of the 9th International Conference on Natural Computation (ICNC ‹13), pp. 777–781, IEEE, Shenyang, China, July 2013.
28. G. Li, J. Feng, C. Wang, and J. Wang, "Hardware/software partitioning algorithm based on the combination of genetic algorithm and Tabu search," Engineering Review, vol. 34, no. 2, pp. 151–160, 2014. ·
29. T. Eimuri and S. Salehi, "Using DPSO and B&B algorithms for Hardware/Software partitioning in co-design," in Proceedings of the 2nd International Conference on Computer Research and Development (ICCRD ‹10), pp. 416–420, May 2010.
30. H. Han, W. Liu, W. Jigang, and G. Jiang, "Efficient algorithm for hardware/software partitioning and scheduling on MPSoC," Journal of Computers (Finland), vol. 8, no. 1, pp. 61–68, 2013.·
31. M. Mehendale, S. Das, M. Sharma et al., "A true multistandard, programmable, low-power, full HD video-codec engine for smartphone SoC," in Proceedings of the IEEE International Solid-State Circuits Conference Digest of Technical Papers (ISSCC ‹12), pp. 226–228, San Francisco, Calif, USA, February 2012.
32. J. Henkel, "Low power hardware/software partitioning approach for core-based embedded systems," in Proceedings of the 36th Annual Design Automation Conference (DAC ‹99), pp. 122–127, New Orleans, La, USA, June 1999.
33. F. Kriebel, S. Rehman, D. Sun, M. Shafique, and J. Henkel, "ASER: adaptive soft error resilience for reliability-heterogeneous processors

in the dark silicon era," in Proceedings of the 51st Annual Design Automation Conference (DAC ‹14), ACM, San Francisco, Calif, USA, June 2014.

34. B. Raghunathan, Y. Turakhia, S. Garg, and D. Marculescu, "Cherry-picking: exploiting process variations in dark-silicon homogeneous chip multi-processors," in Proceedings of the 16th Design, Automation and Test in Europe Conference and Exhibition (DATE ‹13), pp. 39–44, EDA Consortium, Grenoble, France, March 2013.
35. M. Shafique, S. Garg, J. Henkel, and D. Marculescu, "The EDA challenges in the dark silicon era," in Proceedings of the 51st Annual Design Automation Conference (DAC ‹14), San Francisco, Calif, USA, June 2014.
36. J. Henkel, L. Bauer, N. Dutt et al., "Reliable on-chip systems in the nano-era: lessons learnt and future trends," in Proceedings of the 50th Annual Design Automation Conference (DAC ‹13), ACM, Austin, Tex, USA, June 2013.
37. J. Henkel, L. Bauer, H. Zhang, S. Rehman, and M. Shafique, "Multi-layer dependability: from microarchitecture to application level," in Proceedings of the 51st Annual Design Automation Conference (DAC ‹14), San Francisco, Calif, USA, June 2014.
38. H. Kaul, M. Anders, S. Hsu, A. Agarwal, R. Krishnamurthy, and S. Borkar, "Near-threshold voltage (NTV) design: opportunities and challenges," in Proceedings of the 49th Annual Design Automation Conference (DAC ‹12), pp. 1153–1158, San Francisco, Calif, USA, June 2012.
39. J. Cong, M. A. Ghodrat, M. Gill, B. Grigorian, and G. Reinman, "Architecture support for accelerator-rich CMPs," in Proceedings of the 49th Annual Design Automation Conference (DAC ‹12), pp. 843–849, ACM, San Francisco, Calif, USA, June 2012.
40. M. J. Lyons, M. Hempstead, G.-Y. Wei, and D. Brooks, "The accelerator store: a shared memory framework for accelerator-based systems," Transactions on Architecture and Code Optimization, vol. 8, no. 4, article no. 48, 2012.
41. J. Allred, S. Roy, and K. Chakraborty, "Designing for dark silicon: a methodological perspective on energy efficient systems," in Proceedings of the ACM/IEEE International Symposium on Low Power Electronics and Design (ISLPED ‹12), pp. 255–260, Redondo Beach, Calif, USA, August 2012.

42. H. Esmaeilzadeh, E. Blem, R. St. Amant, K. Sankaralingam, and D. Burger, "Dark silicon and the end of multicore scaling," in Proceedings of the 38th Annual International Symposium on Computer Architecture (ISCA ‹11), pp. 365–376, San Jose, Calif, USA, June 2011.

43. Y. Turakhia, B. Raghunathan, S. Garg, and D. Marculescu, "HaDeS: architectural synthesis for heterogeneous dark silicon chip multi-processors," in Proceedings of the 50th Annual Design Automation Conference (DAC ‹13), Austin, Tex, USA, June 2013.

44. M. Shafique, S. Garg, T. Mitra, S. Parameswaran, and J. Henkel, "Dark silicon as a challenge for hardware/software co-design," in Proceedings of the International Conference on Hardware/Software Codesign and System Synthesis (CODES ‹14), ACM, New Delhi, India, October 2014.·

45. D. Diamantopoulos, S. Xydis, K. Siozios, and D. Soudris, "Mitigating memory-induced dark silicon in many-accelerator architectures," IEEE Computer Architecture Letters, vol. 14, no. 2, pp. 136–139, 2015.

46. P. Mantovani, E. G. Cota, K. Tien et al., "An FPGA-based infrastructure for fine-grained DVFS analysis in high-performance embedded systems," in Proceedings of the 53rd Annual ACM IEEE Design Automation Conference (DAC ‹16), Austin, Tex, USA, June 2016.

47. P. V. Knudsen and J. Madsen, "PACE: a dynamic programming algorithm for hardware/software partitioning," in Proceedings of the 4th International Workshop on Hardware/Software Co-Design (Codes/CASHE ‹96), pp. 85–92, IEEE, Pittsburgh, PA, USA, March 1996.

48. G. Stitt, F. Vahid, G. McGregor, and B. Einloth, "Hardware/software partitioning of software binaries: a case study of H. 264 decoder," in Proceedings of the IEEE/ACM International Conference on Hardware/Software Codesign and System Synthesis, pp. 285–290, New York, NY, USA, 2005.

49. J. R. Armstrong and P. J. M. B. Adhipathi Jr., "Model and synthesis directed task assignment for systems on a chip," in Proceedings of the 15th International Conference on Parallel and Distributed

Computing Systems, pp. 472–475, Cambridge, Mass, USA, 2002.

50. K. K. Parhi and T. Nishitami, Digital Signal Processing for Multimedia Systems, CRC Press, Boca Raton, Fla, USA, 1999.
51. R. Ayadi, B. Ouni, and A. Mtibaa, "A partitioning methodology that optimizes the communication cost for reconfigurable computing systems," International Journal of Automation and Computing, vol. 9, no. 3, pp. 280–287, 2012.
52. M. Jemai, S. Dimassi, B. Ouni et al., "Combined partitioning hardware-software algorithms," International Journal of Computer Applications, vol. 119, no. 4, pp. 11–15, 2015.
53. B. C. Sahoo, Design and power estimation of booth multiplier using different adder architectures [Ph.D. thesis], National Institute of Technology, Rourkela, India, 2013.
54. B. A. B. Sarif, M. Pourazad, P. Nasiopoulos, and V. C. M. Leung, "A study on the power consumption of H.264/AVC-based video sensor network," International Journal of Distributed Sensor Networks, vol. 2015, Article ID 304787, 10 pages, 2015.
55. S. Banerjee, E. Bozorgzadeh, and N. Dutt, "Physically-aware HW-SW partitioning for reconfigurable architectures with partial dynamic reconfiguration," in Proceedings of the 42nd Design Automation Conference (DAC ‹05), pp. 335–340, ACM, Anaheim, Calif, USA, June 2005.

CHAPTER 9

AN ANALYSIS OF ROBUSTNESS APPROACHES FOR THE AIRPORT BAGGAGE SORTING STATION ASSIGNMENT PROBLEM

Amadeo Ascó

School of Computer Science, University of Nottingham, Nottingham NG8 1BB, UK

ABSTRACT

In the Airport Baggage Sorting Station Assignment Problem (ABSSAP), the Baggage Sorting Stations (BSSs) are assigned to flights for the period of time necessary to perform their service for a given flights' schedule. But the flights schedule may change on the day of operation which may deem the original assignment of some flights to BSSs infeasible. These changes may create conflicts between those flights whose schedules have changed and may not be restricted to those flights but propagating to the other flights for different reasons. Conflicts depend on the original assignments for the real arrival and departure flight times on the day of operation. It is therefore desirable to consider potential delays on the day of operation when generating the original flight assignments to BSSs, such that the final flight assignments differ little or do not differ at all from the original assignments on the day of operation. The term robustness is here used to give an indication of the degree to which this has been achieved. Some existing approaches originally presented in the Airport Gate Assignment Problem (AGAP) are adapted to the ABSSAP, other approaches are suggested for generating assignments which take account of potential perturbations on the day of operation for the ABSSAP, and all of them are then compared. It is shown that the suggested approaches by themselves do not perform better than the other considered approaches but when combined they enhance the result further compared to when each approach is used alone.

INTRODUCTION

Flight delays are caused by many factors like airport security, weather conditions, unavailability of required resources (mechanical breakdown), delayed propagation, airport congestion, etc.

The ABSSAP corresponds to the assignments of flights to BSSs, whereas the BSSs are the end elements in the baggage system at the ground of the airport terminal where the baggage is collected from and sorted by the baggage handlers or to where the baggage from arrived flights are placed to enter the baggage system. Figure 1shows an overall simple view of the airport baggage handling system.

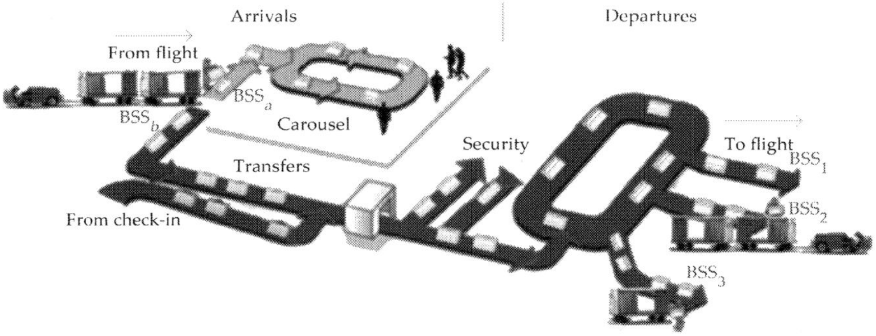

Figure 1: Simple view of an airport baggage handling system with the BSSs.

Once the baggage has been loaded into the aircraft hold, the BSSs assigned to that flight are ready for use on the next scheduled flights, and any delay on a flight at the gate will not necessarily have repercussions on these. However, such delays may affect the BSSs assigned to the flight scheduled for that gate next if such a flight is held waiting for the gate to become free. BSSs will then need to hold the baggage longer, thus potentially affecting the following assignment in turn.

Delays on the day of operation may render some assignments infeasible which need to be reassigned. It is therefore desirable to account for potential delays on the day of operation when generating the flight assignments to BSSs in the planning stage, such that the final flight assignments differ little or do not differ at all from the original assignments on the day of operation. The degree to which this is achieved is an indication of the solution robustness, so a solution which requires fewer reassignments is said to be more robust than those solutions requiring more reassignments. Robustness is the ability of assignments to resist changes consequence of perturbations by reducing or removing the need to reassign current assignments. Mulvey et al. [1] characterize the desirable properties of a solution to model robustness.

There are different ways of increasing robustness depending on the intended effect. One of the most simple and widely used methods is the introduction of a buffer time between assignments which allows the absorption of small disturbances; implementation of this approach for the ABSSAP is presented in [2–5]. The amount of buffer time may take different contributory factors into account, which could perturb the schedules of handlers, airline, airport origin, destination, and flight. The size of the flight is normally related to the travel distance, longer distances presenting a higher probability of disturbances, which may accumulate generating a higher level of delays. Shorter distances present less chance of disturbances. The location of the origin and destination airports has a direct effect on potential disruptions, given that they place a constraint on the permitted routes possible and certain circumstances applicable to them, such as weather patterns during summer and winter or on the equator. Whereas the buffer time approach considers a predefined period of time, where its reduction is considered as a measure of robustness, the "idle time" refers to the time between two consecutive assignments to the same BSS, from the end time of one activity to the base starting time of the following assignment, also called gap. Other approaches for improving the robustness make use of the distribution of "idle time" and the reduction of the number of reassignments of the disrupted schedules.

Buffer time has been used in the scheduling of BSSs and airport gates, which may be applied between two consecutive flights to the same BSS in order to absorb small disturbances in the real system behavior and was studied in [1, 6]. It was used in the ABSSAP in [2, 3, 5, 7] where it was used in the optimization of the aircraft turnaround process. The AGAP has some characteristics similar to the ABSSAP, and buffer times have been commonly considered for the AGAP in [7–11]. Yan and Huo [12] provided a sensitivity analysis for the AGAP buffer time, noting that the length of buffer time significantly influences the gate assignment process, so a reasonable minimum value should be used. Yan et al. [8] looked at the suitability of Flexible Buffer Times (FBTs) where, given low delays, short FBTs usually improve real-time objectives, such as the reassigning of an incoming aircraft at a minimum distance. Wei and Liu [13] showed the feasibility and effectiveness of using a fuzzy model in conjunction with fixed buffer times for the AGAPs. Ascó et al. [4] used buffer times to cope with small perturbations in the ABSSAP, and several constructive algorithms were also studied. Wu and Caves [7, 14] showed the significance of a correct use of scheduled buffer time in maintaining schedule punctuality and performance by balancing the trade-offs between schedule punctuality and aircraft utilization.

PROBLEM UNDER CONSIDERATION

The ABSSAP considered in this paper may be summarized as the assignment of available BSSs to flights which have already been scheduled. In the ABSSAP, the flights will already have been assigned to stands, which are often grouped along piers around the terminals, and there will usually be some bias in this allocation, such as to airline preferences. This problem was described and studied in [2–5]. In this paper, we concentrate on finding solutions that comply with the constraints and take into account the objectives concentrating on the performance of different robustness approaches described in Section 3.

Overall Problem

Aircraft are usually parked at their allocated stand, around an airport terminal. Two layouts are considered in this paper which were presented in [2].

A list of some of the constants for this problem is presented in Table 1 and a list of the decision variables is shown in Table 2.

Table 1: The constants for the model

Constant	Description
N	Denotes the number of BSSs.
M	Denotes the number of flights.
P_j	The total number of activities to be serviced by baggage service stations for a given flight j, which also equates to the total number of sorting stations required to fully service flight j, $P_j > 0$.
i	Denotes a BSS where BSS 0 represents the dummy sorting station to which flights are assigned, if they cannot be assigned to a real BSS, $i \in \{0, \ldots, N\}$.
j	Denotes a flight, $j \in \{1, \ldots, M\}$.
T_j	Denotes the required service time for flight j (1 hour for short haul and 1.33 hours for long haul).
B_j	Denotes the desired buffer time for flight j the time for which its BSS should be idle prior to this flight being serviced. B_j is considered to be 15 minutes for short haul and 30 minutes for long haul flights.
e_j	Denotes the end service time for flight j.
t_j	Denotes the target starting service time for flight j ($t_j = e_j - T_j - B_j$).
τ_j	Denotes the the base starting service time for flight j ($\tau_j = e_j - T_j$).

Table 2: List of the decision variables used in this ABSSAPs' model

Name	Description
y_{ijp}	Specifies the assignment of flights to sorting stations. $y_{ijp} = 1$ if baggage sorting station $i \in [1, \ldots, N]$ is allocated to flight $j \in [1, \ldots, M]$ for $p \in [1, \ldots, P_j]$ and 0 otherwise. If each flight only requires one activity, which means that each flight only requires one BSS, then this variable can be expressed as y_{ij}.
r_{jp}	Specifies the necessary reduction in service time for activity $p \in [1, \ldots, P_j]$ of flight $j \in [1, \ldots, M]$, given the service starting time allocated, s_{jp}.
s_{jp}	The service starting time allocated to activity $p \in [1, \ldots, P_j]$ of flight $j \in [1, \ldots, M]$ and given that a sorting station can only service one flight at a time. s_{jp} can be determined from r_{jp} since $s_{jp} = t_j - r_{jp}$.

The aim is to find y_{ijp} values such that the objective function (3) is maximized, subject to the constraints expressed by (1) and (2) and the different robustness approaches studied in this paper which are presented in Section 3.

$$\sum_{i=0}^{N} y_{ijp} = P_j \quad \forall j \in \{1, \ldots, M\}, \ \forall p \in \left\{1, \ldots, P_j\right\}. \tag{1}$$

For any pair of different flights where service times overlap, if the overlap in service times is greater than the maximum reduction allowed (B_{lq} for activity q of flight l), then both flight activities cannot be assigned to the same BSS. Thus, inequality (2) applies to any such pair of flights, j and l (j ¹ l), where $t_{lq} < e_j \leq e_l$ and $(e_j - t_{lq}) > B_{lq}$

$$y_{ijp} + y_{ilp} \leq 1. \tag{2}$$

Equation (1) states that each flight can be assigned either to exactly one real BSS or to the dummy sorting station. Inequality (2) states that flights cannot be assigned to the same real BSS if their service times overlap. If all flights only require one BSS, then the variables are reduced to y_{ij}, r_j, and s_j.

Although this is inherently a multiobjective problem, the importance of ensuring maximal assignment of flights to BSSs (top priority) and the relative importance of obtained robust solutions (second priority) allow these objectives to be combined into a single compound objective (3) with weights W_1, W_2, and W_3 chosen to implement these priorities.

$$\underbrace{W_1 * \sum_{i=1}^{N}\sum_{j=1}^{M}\sum_{p=1}^{P_j} y_{ijp}}_{\text{assignment objective}} - \underbrace{W_2 * \text{Robustness}}_{\text{robustness objective}}$$

$$\underbrace{-W_3 * \sum_{j=1}^{M}\sum_{p=1}^{P_j}\left(C_j \cdot \sum_{i=1}^{N} y_{ijp} \cdot d_{ij}\right)}_{\text{distance objective}}. \qquad (3)$$

The first element in (3) aims to maximize the assignment of flights to BSSs, the second aims to increase the robustness, and the third aims to minimize some distance cost associated with the assignments, where C_j is a factor related to the amount of baggage for flight j (assumed to be 1 in all cases for this paper) and d_{ij} is a measure of the distance or cost incurred from assigning BSS i to flight j. This aims to ensure that flights are allocated to appropriate BSSs.

The following two points were originally defined in [4] and will be observed to be useful later when defining the new suggested robustness approaches and when interpreting the results for the ABSSAP.

The Lower Maximum Assignment Point (LMAP) is the minimum number of resources required to service a certain number of activities when the service starting time (s_{jp}) coincides with the base starting service time (τ_j)

The Upper Maximum Assignment Point (UMAP) is the minimum number of resources required to service a certain number of activities when the service starting time (s_{jp}) coincides with the target starting service time (t_{jp}).

Two examples of both points are shown in Figure 2, where the maximum values of each diagram correspond to the LMAP and UMAP.

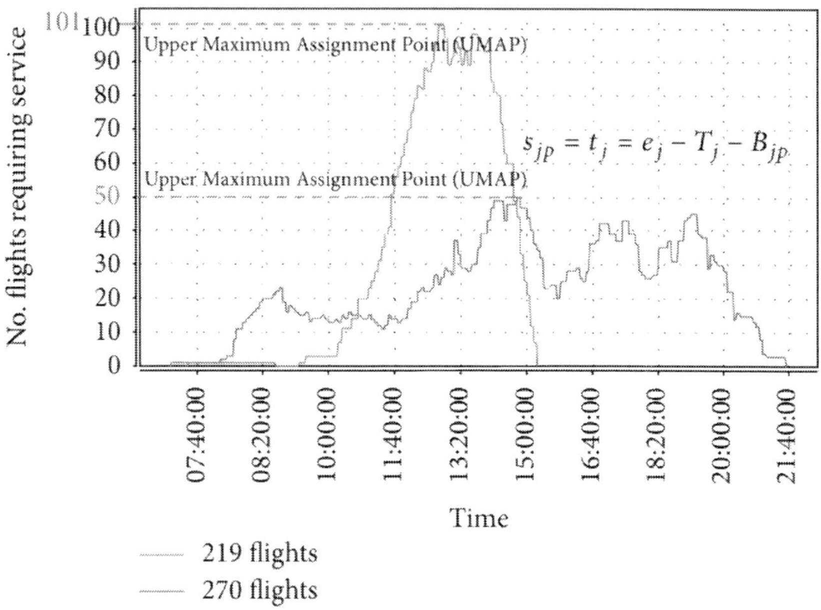

(a)

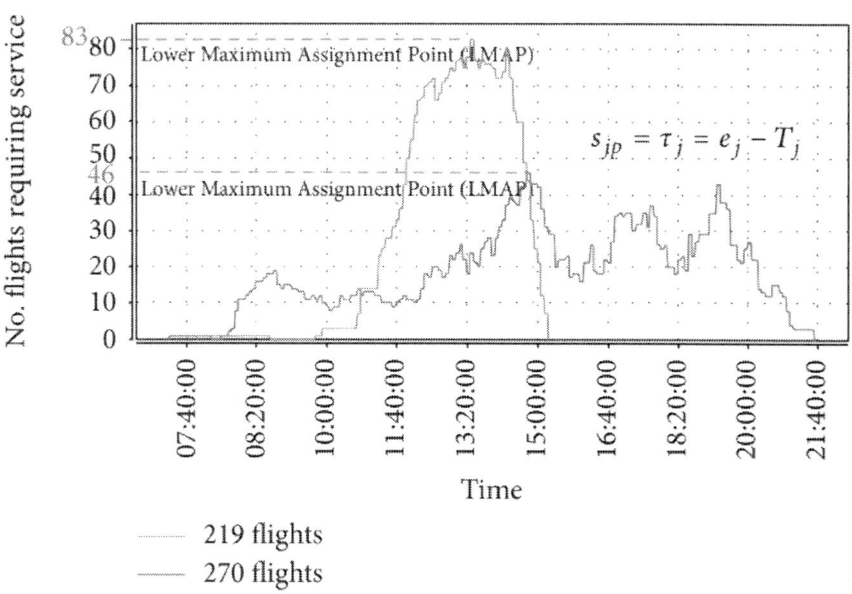

(b)

Figure 2: Flights service distribution for the two days' data sets obtained from the website of British Airports Authority (BAA).

Robustness

A scheduled assignment is said to be in conflict if the completion of its service time is greater than the commencement of the service time of the next assignment at the same BSS. When a delayed flight affects the assignment of subsequent flights to the BSS, then there are two ways it can be corrected: either reassign the conflicting flight or reassign the subsequent flight to the conflicting one. When reassigning a conflicting flight or subsequent flights to another BSS, a situation may arise where the reassigned flight is in conflict with the subsequent flight at the new BSS. Some reassignment may therefore have a downstream effect on the overall schedule, producing further conflicting flights requiring further reassignments, thus potentially increasing the problem difficulty later on.

In the ABSSAP, those flights which are late in arriving at their assigned stand are considered to be a perturbation, since baggage cannot be loaded into the aircraft at the scheduled time, and it needs to be held longer at the BSS. Any extension of an aircraft's stay on its assigned gate should not have an effect on its assigned BSS, as the baggage should have already been loaded into the aircraft, such that the BSS is free for use in its next assignment. This means that not all aircraft delays will affect their assigned BSS.

The main objective of this is to reduce the number of BSSs which have to be reassigned on the day the schedule is put into practice. In the ABSSAP, e_j is the end of the service time of flight j, τ_j is the base starting service time of flight j, and the variable y_{ij} has a value of 1 if flight j is assigned to BSS i, or y_{ij} is zero otherwise, as described in Section 2.1. A new decision variable x_{jk} is introduced with a value of 1 if flights j and k are assigned to the same BSS (i.e., if $y_{ij} = y_{ik} = 1$ for $i \in [1, \ldots, N]$ and $j, k \in [1, \ldots, M]$), or xjk is 0 otherwise. On the day the schedule is implemented, the real times for the flights j and k correspond to e'_j, τ'_j, e'_k, and τ'_k for flights j and k, respectively, where $\tau'_j \geq \tau j$ and $\tau'_k \geq \tau k$, and the flights ordered by their base starting times, for j τ'_k and $\tau'_j < e'_k$, as shown in Figure 3. Conflicts depend on the original assignments for the real arrival and departure flight times on the day of operation. It is therefore desirable to consider potential delays on the day of operation when generating the original flight assignments to BSSs, such that the final flight assignments differ little or do not differ at all from the original assignments on the day of operation. B_j refers to the buffer time originally assigned to the flight j, which may be absorbed by perturbations without raising the need for a reassignment. But overlapping, $\tau'_k < e'$, will deem the assignment infeasible and require reassigning it.

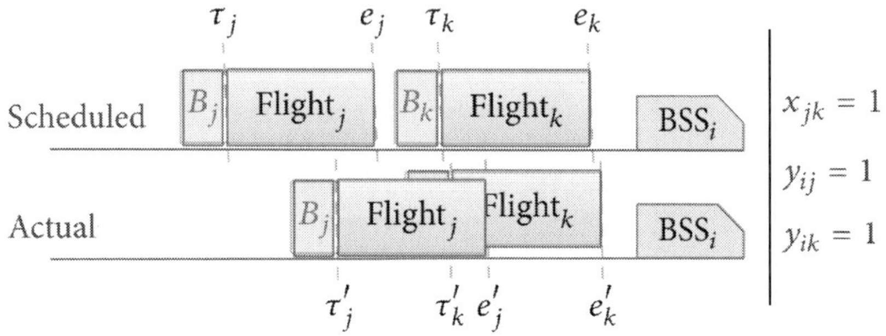

Figure 3: Example of conflict between two flights originally assigned to the same BSS.

ROBUSTNESS APPROACHES

In this section, some robustness approaches are presented and others are suggested. These are then studied and compared in the following section.

Minimize Reduction in Service Time

It may be possible to gain robustness by reordering assignments between BSSs so that "idle time" between flights consecutively assigned to the same BSS is greater, as shown in Figure 4.

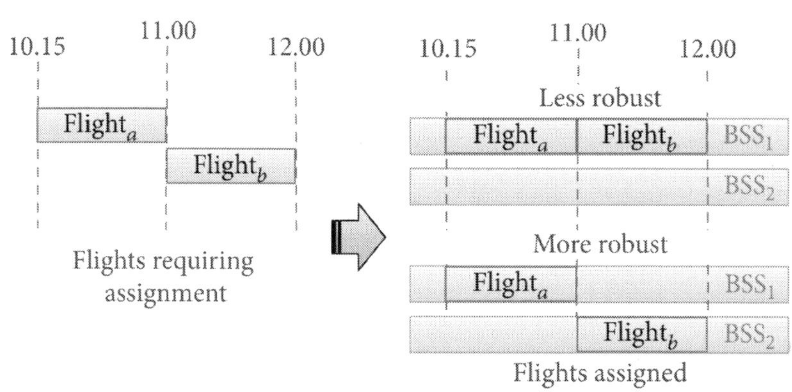

Figure 4: Simple example of two schedules, with the same flights, where one is obviously more robust in respect of perturbations than the other.

Figure 4 shows two potential solutions with different robustness. While any delay to flight "*a*" in the "less robust" solution will certainly affect flight "*b*," which will in turn have to be reassigned to another BSS, in the "more robust" solution a delay in flight "*a*" will not affect flight "*b*." So the "more robust" solution is preferable to that of the "less robust" solution.

Mangoubi and Mathaisel [15] proposed the use of "buffer times" between two flights which are consecutively assigned to the same gate in order to obtain robust assignments, defining the reduction in service as that part of buffer time which overlaps with the previous assignment to the same gate. Given the detrimental effects that the reduction in service time has on the robustness of the assignment as against real-life delays, it is advisable to minimize the total reduction in service time, thus maximizing buffer times. This objective, total reduction in service time (TRS), can be expressed by formula (4).

$$\min \sum_{j=1}^{M} \sum_{p=1}^{P_j} r_{jp}. \tag{4}$$

If a fitness function is defined as a weighted sum of the different objectives, as used here, and the robustness uses the reduction in service time only for assigned flights, then particular weights may compromise the importance of the main objective (maximization of the number of assignments), where solutions with lower numbers of assignments are favored over those with higher numbers of assignments because of the robustness objective.

The use of formula (4) treats any reduction in service equally, so it does not make a distinction between reducing all the assignment buffer time and allocating it to another flight from another solution where both flights share the available "idle time," as shown in Figure 5. However, flight 3 with maximum reduction in service in solution "b" (Figure 5) will be unable to absorb any delay on the day of operation, although the same flight in the alternative solution "a" (Figure 5) will be able to do so, making it the preferable choice.

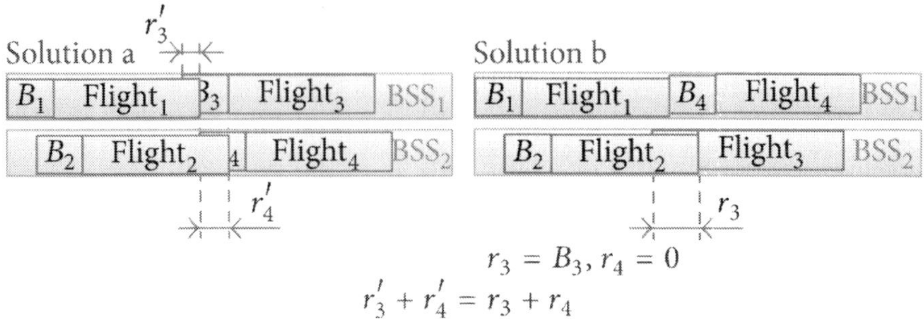

Figure 5: Simple example of a more robust schedule of four flights and two sorting stations using the reduction in buffer time.

A nonlinear penalty function is required to take account of this, such as formulas (5) and (6).

$$u_{ijp} = \begin{cases} \arctan\left(\frac{\tau_j - e_k}{B_j}\right) - \frac{\Pi}{2} & \text{if } k < j,\ y_{ijp} = y_{ikq} = 1,\ \sum_{l=k+1}^{j} \sum_{p=1}^{P_l} y_{ilp} = 1, \\ 0 & \text{otherwise,} \end{cases} \tag{5}$$

$$\max \sum_{i=1}^{N} \sum_{j=1}^{M-1} \sum_{p=1}^{P_j} u_{ijp}. \tag{6}$$

Formula (5) defines the penalty u_{ijp} for assigning activity p of flight j to sorting station i, with the total penalty being represented by formula (6). There is only a penalty for consecutive assignments to the same sorting station, between the assignment and the previous assignment to the same sorting station. The reason for using arctangent is based on the properties of this function which imposes stronger penalties around the point at which the flights are assigned service time without any buffer time to the point at which all the buffer time is retained (by dividing the member of arctangent by B_k). The effect is reduced as the time separation between the assignments increases, but the contribution of any time separation is always considered negative. The constant $\pi/2$ is used so that there is always a cost associated with assigning two flights to the same sorting station; otherwise it would be a benefit. Considering the arctangent for each flight increases the individual effect on the overall objective, in contrast to using the arctangent of the sum of all of the flight contributions.

The robustness objective represented by formula (6) also takes account of the objective of "minimizing the service reduction," as being the sum of the reduction in buffer time of each flight assigned.

Distribute Idle Time

Bolat [16] proposed the distribution of "idle time" uniformly amongst gates for the AGAP. In the case of the ABSSAP, the distribution of the "idle time" uniformly amongst BSSs can be considered, where "idle time" is calculated as the time between the start of service time on a flight's assignment to BSS and the end of service time of the flight assigned immediately before the same baggage sorting station, as shown in formula (7). Whereas the buffer time implies preference for a particular gap size between consecutive assignments to the same sorting station, an "idle time" does not. When using the "idle time" it is normally intended that the gap size should be as large as possible. The reason for this is to increase the probability that, even with a delay, completion of service time for flight presently assigned will still be earlier than the start of service time for the next flight assigned to the same BSS.

$$\Delta t_j = (s_j - e_k)$$

$$\text{for } j < k,\ y_{ijp} = y_{ikq} = 1 \text{ where } \nexists l \in [k,\ldots,M] \text{ with } r_l < e_k;\ j,k \in [1,\ldots,M],\ p \in [1,\ldots,P_j],\ q \in [1,\ldots,Pk],\ i \in [1,\ldots,N_i], \tag{7}$$

$$v^+ = \max\{\Delta t_j \mid j \in [1,\ldots,M]\}, \tag{8}$$

$$v^- = \min\{\Delta t_j \mid j \in [1,\ldots,M]\}, \tag{9}$$

$$\min(v^+ - v^-). \tag{10}$$

Formulas (8) and (9) refer to the maximum and minimum "idle time" for a solution, respectively, while formula (10) represents the objective as the difference between both the maximum (V^+) and the minimum (V^-) "idle times" (Δt_j, in (7)) for the same solution. Figure 6 shows a simple example of the robustness of two solutions, where solution "a" is more robust than solution "b." In solution "a" the delay of flight 3 will not affect flight 4 and needs to be considerably larger to affect flight 5, whereas in solution "b" small delays in flight 3 will affect flight 4.

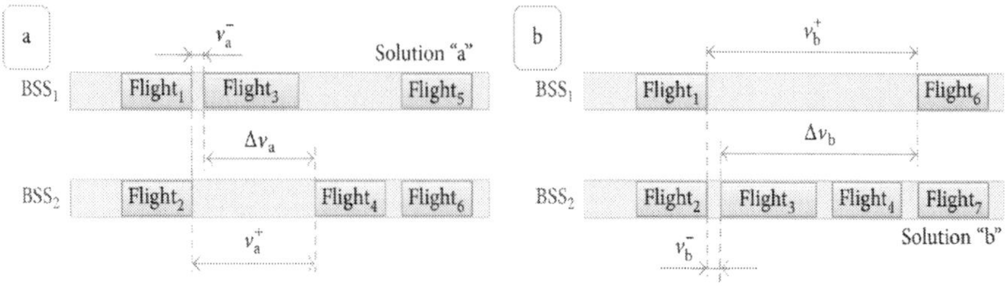

Figure 6: Examples of a range of idle times for different solutions, where solution "" is clearly better than solution "" for formula (10).

Formula (10) assumes that all flights within a solution have been assigned, which may not be the case in certain circumstances, where the maximum possible assignment is lower than a full assignment (simulation) or where the initial solution(s) has some flights remaining unassigned, and an example of this is shown in Figure 7(a). It should also be observed that this objective may conflict with the maximum assignment objective (first objective), as shown in Figure 7(b), where solution "f" has a smaller ΔV compared to solution "e," which is based on formula (10). This means that solution "f" is considered more robust, but solution "e" would be preferable because it achieves more assignments. Thus, it will be necessary to select the objective weights appropriately, where the fitness is a weighted sum of the different objectives, in order to ensure the correct selection of the solution.

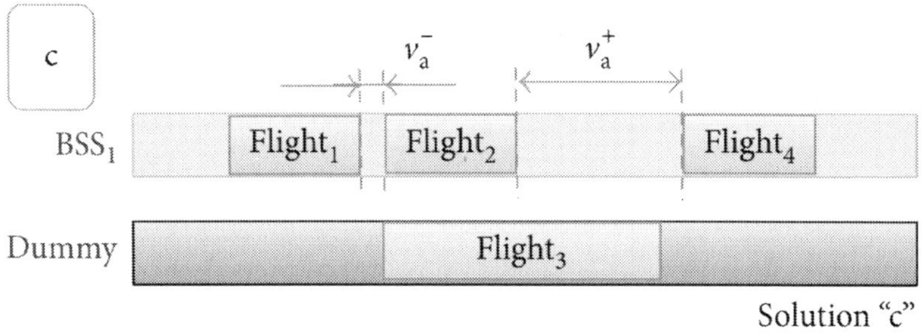

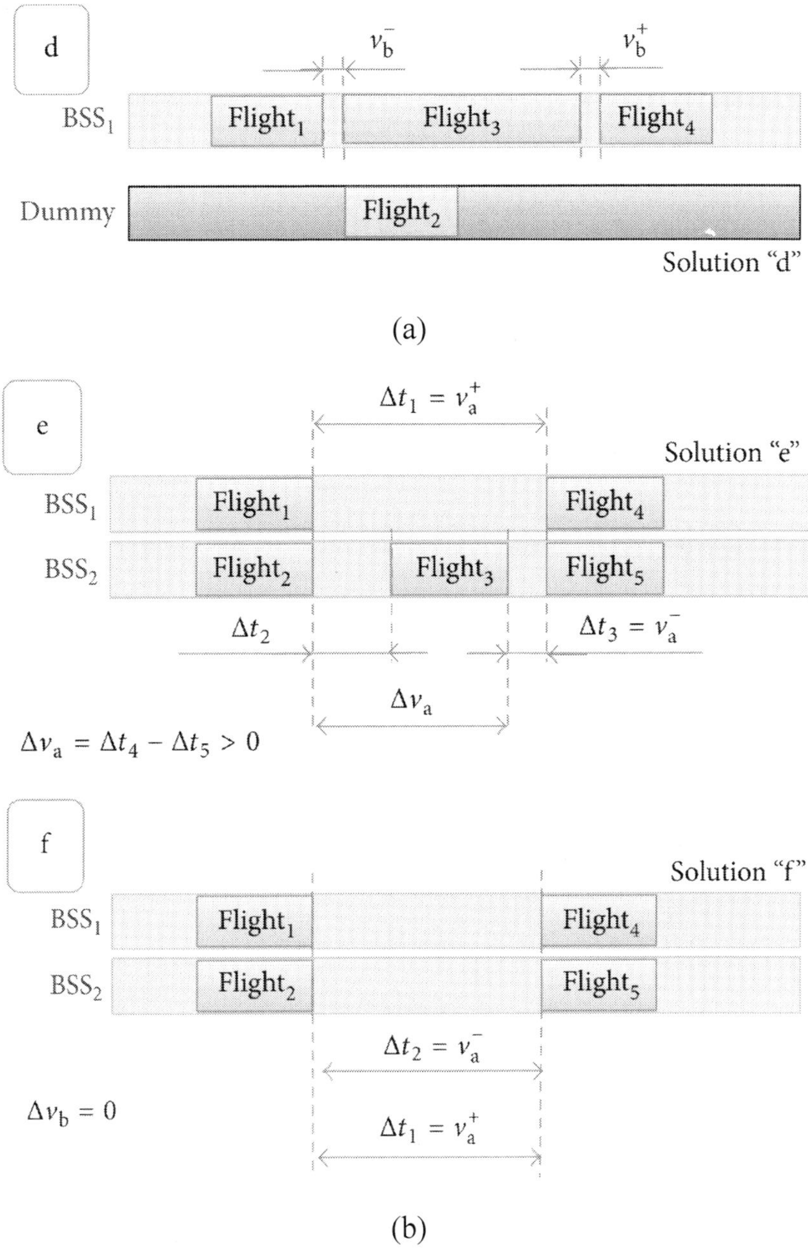

Figure 7: Examples of a range of idle times for different solutions using formula (10).

Bolat [9, 17] extended the model by minimizing the variance of the idle times.

$$\min \sqrt{\sum_{j=1}^{M}\left(\Delta t_j - \underbrace{\frac{\sum_{j=1}^{M}\Delta t_j}{M}}_{\text{mean "idle time"}}\right)^2} \qquad (11)$$

Reduce Reassignment on Disruption

The ability to reassign all flights directly affected by a disruption is desirable, without the need to reassign other flights. The intention here is to generate schedules which take account of this objective, allowing such reassignment to be performed more frequently.

One way to achieve this objective would be to count the number of assignments between which a reassignment could be placed when necessary. Whether the reassigned flights are on the same pier/side, as well as how many reassignments could be absorbed by a pair of assigned flights, must all be taken into account. Figure 8 shows the ability of the "idle time" between the two flights 1 and 8 to accommodate flights 3, 4, and 5, should one of them be delayed. Its reassignment to BSS_1 may be sufficient, thus avoiding transfer of any delay to other assignments.

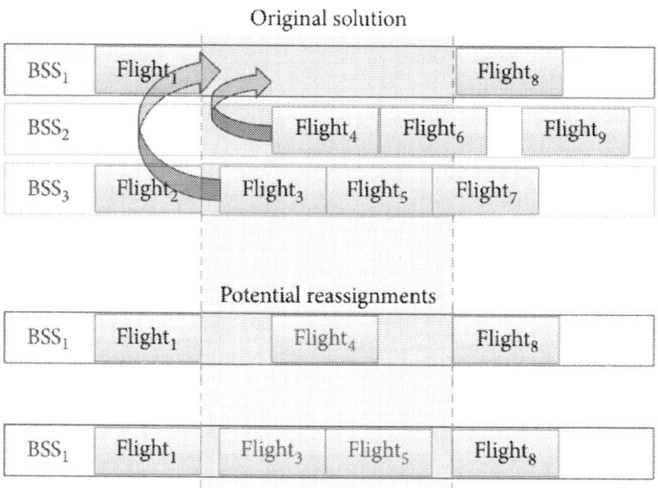

Figure 8: An example of the capacity to absorb reassignments.

The following model is proposed where the capacity to absorb reassignments may be achieved by weighting each reassignment by the Inverse of 1 plus the distance between the BSSs (d'_{in}, presented in Section 2.1), given that such distance (d'_{in}) may be zero, where all flights are ordered by their base start time (τ_j), as shown in (12) and (13).

The intention is to use the number of flights which could be reassigned between two flights already consecutively assigned to the same BSS, without the need to reassign either of these flights in order to achieve this. Equation (12) states that φ_{jkl} is equal to 1 if flight k's base service duration does not overlap with the base service duration of both flights j and l ($j<l$), both of which are assigned to BSSi, and there is no other flight r between these ($j<r<l$) already assigned to the same BSS as flight j; otherwise φ_{jkl} is zero, as shown in

$$\varphi_{jkl} = \begin{cases} 1 & \text{if } j < k < l,\ e_j \le \tau_k,\ e_k \le \tau_l\ \forall j,k,l \in [1,\dots,M]\ \nexists r \in (j,\dots,l), \\ & \text{with } y_{ijp} = y_{irq} = 1 \text{ for any } p \in \left[1,\dots,P_j\right],\ q \in \left[1,\dots,P_r\right], \\ & y_{ikz} = 0\ \forall z \in \left[1,\dots,P_k\right], \\ 0 & \text{otherwise.} \end{cases} \tag{12}$$

The objective is to maximize formula (13), which weights the contribution of each potential reassignment, based on which BSS the reassigned flight was originally assigned to. The underlying idea is that closer reassignments are preferred to more distant ones, but the ability to reassign without affecting other assignments is preferable.

$$\sum_{i=1}^{N}\sum_{j=1}^{M}\sum_{p=1}^{P_j}\sum_{l=j+1}^{M}\sum_{q=1}^{P_l}\sum_{k=j+1}^{l-1} y_{ijp} * y_{ilq} * \varphi_{ikl} * \underbrace{\sum_{n\neq i}^{N}\sum_{z=1}^{P_k}\frac{y_{nkz}}{d'_{\text{in}}+1}}_{\text{effect of the BSS assigned}}. \tag{13}$$

The representation in formula (13) also needs to include the border cases relating to the first and last assignments in BSS as shown in Figure 9.

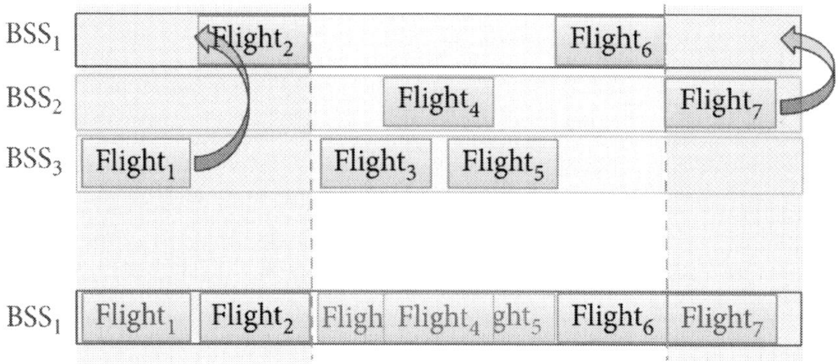

Figure 9: An example of border assignments in baggage sorting stations.

The previous formula could be extended to cover the border cases by assigning two extra dummy flights to all available BSSs; first, j=0, with the end time being the start of the time period studied, and the second flight, j=M+1, with the start time being the completion of the time period studied, which for our time period would be $e_0 = 0$ and $(e_{M+1} - T_{M+1}) = 24$ hr, with $y_{i0} = yi(M+1) = 1\ \forall i \in [1, \ldots, N]$, as shown in

$$\sum_{i=1}^{N}\sum_{n\neq i}^{N}\left(\frac{\sum_{j=0}^{M}\sum_{p=1}^{P_j}\sum_{l=j+1}^{M+1}\sum_{q=1}^{P_l}\sum_{k=j+1}^{l-1}\sum_{z=1}^{P_k} y_{ijp} * y_{ilq} * y_{nkz} * \varphi_{ikl}}{d_{in}^{\mathrm{BSSs}}+1}\right). \tag{14}$$

Area of Reduction in Service (ARS)

Bolat [16] examined the minimization of the range of idle time and the difference between the maximum and minimum idle times for the AGAP, which was later extended in [9] to consider both minimization of the idle time range and minimization of the idle time variance (Section 3.2). However, these do not take account of the influence or effect which the disruptions have on the schedule due to the time of their occurrence, as all reductions in service time are treated as being the same, irrespective of the time period considered in the whole "planned schedule" for the given set of departures. It is anticipated that the more congested time periods in the "planned schedule" will also represent periods where disruptions are more likely to occur and propagate, extending their effect and further increasing costs.

LMAP and UMAP values are an indication of the difficulty of the problem, and these may be obtained from the distribution of the number of flights requiring service over time, as shown in Figures 10 and 11.

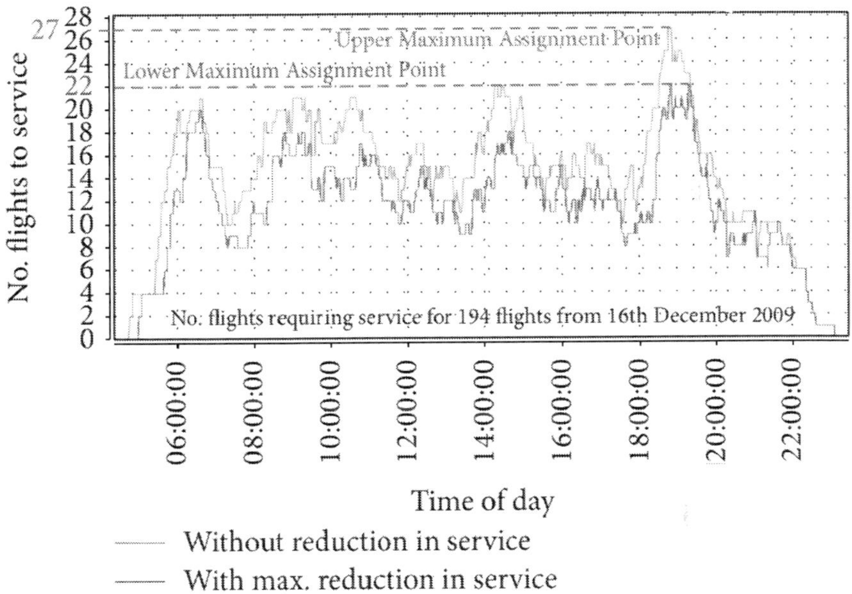

Figure 10: Flight distributions with LMAP and UMAP for 194 flights on 16th December 2009 at Terminal 1 of London Heathrow airport (identified as H1T091216).

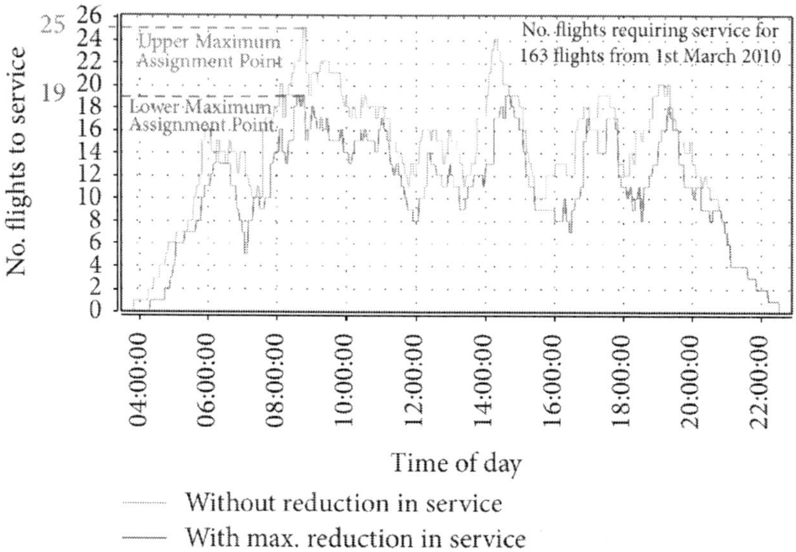

Figure 11: Flight distributions with LMAP and UMAP for 163 flights on 1st March 2010 at Terminal 1 of London Heathrow airport (identified as H1T100301).

Delay during a high flight density period is more likely to propagate given that fewer resources will be available to absorb any reassignment without repercussions on other flights. At the same time, it is these cases where it is most difficult to keep a sufficiently large gap between assignments to the same BSS. In assessing the importance of the time of day when service reduction is performed, it is suggested that it would be advisable to increase penalization of flight assignments with reduced service time when there is a higher flight density. This can be accomplished by calculating the required number of BSSs at different times of the day, as shown in Figures 10 and 11.

The function $f(t)$ refers to the distribution of flights over time with t_s being the schedule starting time and t_e being the schedule end time, as shown in Figure 12. The Average Assignment Point (AAP) is here defined as the number of BSSs for which the distribution of flights would be uniform, which can be calculated by

$$\text{AAP} = \frac{\int_{t_s}^{t_e} f_u(t)\, dt}{t_e - t_s}. \quad (15)$$

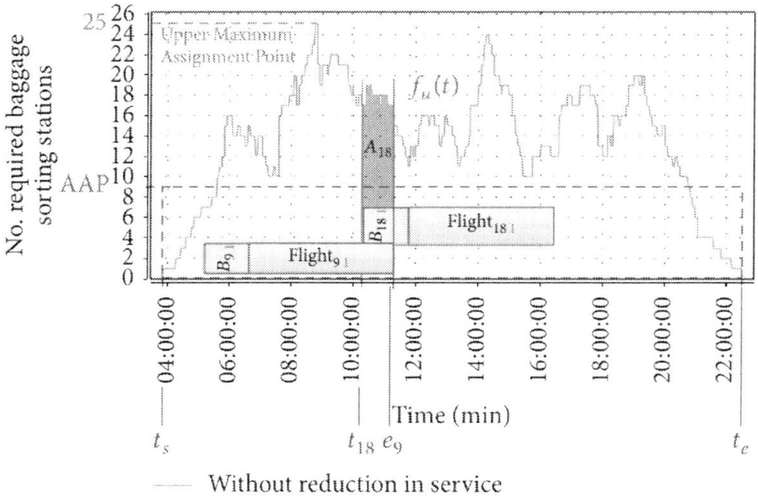

Figure 12: Distribution of flights over time and Area Reduction in Service Objective.

A_j is the density distribution area for the time period from the target service time of flight j (t_j) to the end of service time for the previous flight assigned to the same BSS; for example, in Figure 12 $A_{18} = \int_{t18}^{e9} f_u(t)dt$. AAP$_j$ is here defined as the mean number of flights over the target start time for flight j and the end of service time for the previous flight assigned to the same BSS as flight j; for example, AAP$_{18}$ = $A_{18}/(e_9 - t_{18})$. So the contribution to the objective for assignment j is AAP$_j$/AAP times the reduction in service previously considered in Section 3.1. This corresponds to values greater than one for dense flight regions of the schedule and less than one for underused regions.

The approach proposed intends to penalize more those reductions in flight service time during time periods where more flights require servicing compared to those periods with less service load by means of the AAP and the distribution of flights over time, as shown in (16). The idea is that flights which require servicing during congested periods are more likely to have a knock-on effect compared to those in less congested periods. It is therefore preferable not to reduce the service time of flights at more congested times so much, in order to limit the effect of potential delays:

$$f_2 = -\frac{1}{\text{AAP}} * \sum_{i=1}^{N}\sum_{j=1}^{M}\sum_{p=1}^{P_j} y_{ijp} * A_j. \tag{16}$$

This approach is compared with some of the other approaches described here in Section 4.

Subarea of Reduction in Service (SARS)

In this case, the objective is based on the area between both flight densities "without reduction in service time" and "with reduction in service time" shown in Figure 13, being here called a subarea. The flight density subarea corresponds to the area from the flight target start time (described in Section 2.1) to the previously assigned flight's end service time, which lies between both flight density distributions (f_u and f_l), where a reduction in service time is not permitted and when all the buffer time has been reduced (considering only the base service duration, T).

$$A'_j = \int_{t_j}^{e} \left(f_u(t) - f_l(t)\right) dt. \tag{17}$$

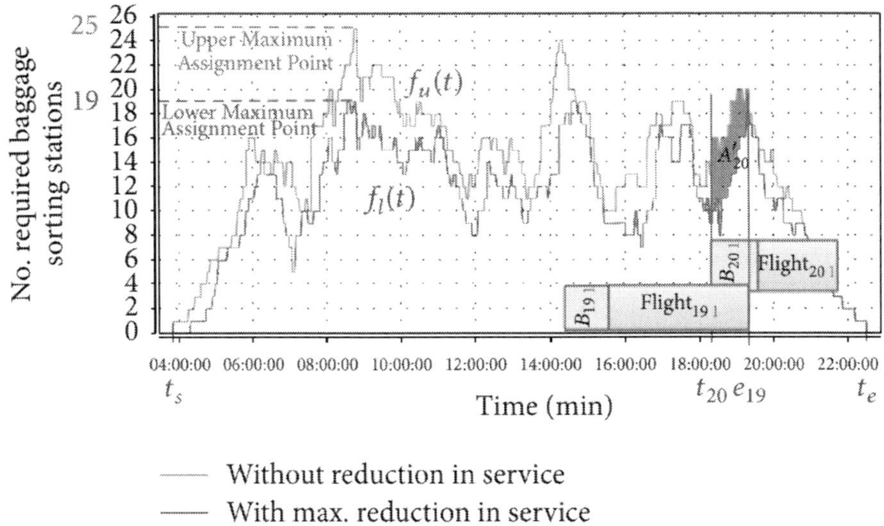

Figure 13: Subarea Reduction in Service Objective.

The approach now proposed has a fitness which covers the subarea divided by the difference between the UMAP and the LMAP for all the assigned flights.

$$f_2 = -\sum_{i=1}^{N}\sum_{j=1}^{M}\sum_{p=1}^{P_j} y_{ijp} * \frac{A'_j}{\text{UMAP} - \text{LMAP}}. \tag{18}$$

Similarly to the Area Reduction in Service (Section 3.4), the main idea is to penalize more heavily those reductions in flight service time which occur in regions with high flight density, as these are more likely to further disrupt the schedule in case of delays. Adding both approaches together with the TRS will be seen in Section 4 to increase the robustness as compared to using each approach individually.

Nevertheless, in cases where the LMAP is equal to the UMAP, the SARS approach cannot be used; an example is shown in Figure 14. In these cases, the area between both flight densities could be used, an approach that is here named Base Subarea Reduction in Service (BSARS). As will be seen in Section 4.1, this approach significantly increases the robustness and widens the range of the quantity of BSSs, in which it performs better when compared to the ARS, TRS, and SARS.

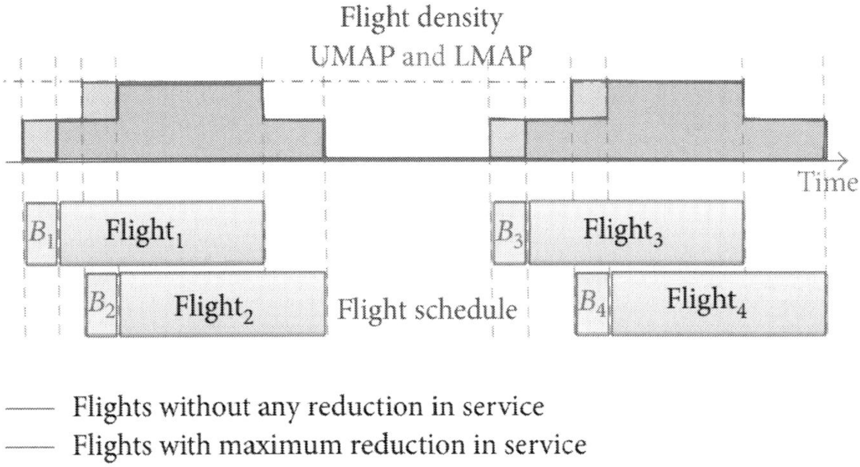

Figure 14: Example of distributions with the same LMAP and UMAP.

These approaches are compared with some of the other approaches introduced in this paper in Section 4.

Unsupervised Estimated Stochastic Reduction in Service (UESRS)

Lim and Wang [18] proposed a stochastic programming model for the AGAP with a robustness cost of conflicts, which is estimated by a function, V(*j*, *k*). Flights are ordered by their base starting service time, so the gap between two flights *j* and *k*, (*j*, *k*), assigned to the same BSS, where *j*< 0, as shown in Figure 15. V(*j*, *k*) is used to estimate the mean conflict probability between flights *j* and *k* assigned to the same BSS, which is a function of the gap *l*(*j*, *k*), where larger gaps between assignments to the same BSSs result in lesser probability of real flight conflicts. V(*j*, *k*)is normalized in (20).

$$l(j,k) = t_k - e_j \quad (t_k = \tau_k - B_k), \tag{19}$$

$$E(p(j,k)) = \sum_{i=1}^{N}\left(y_{ij} * y_{ik} * \frac{v(j,k) - v_{\min}(j,k)}{v_{\max}(j,k) - v_{\min}(j,k)}\right), \tag{20}$$

$$f_2 = -\sum_{j=1}^{M-1}\sum_{k=j+1}^{M} E(p(j,k)). \tag{21}$$

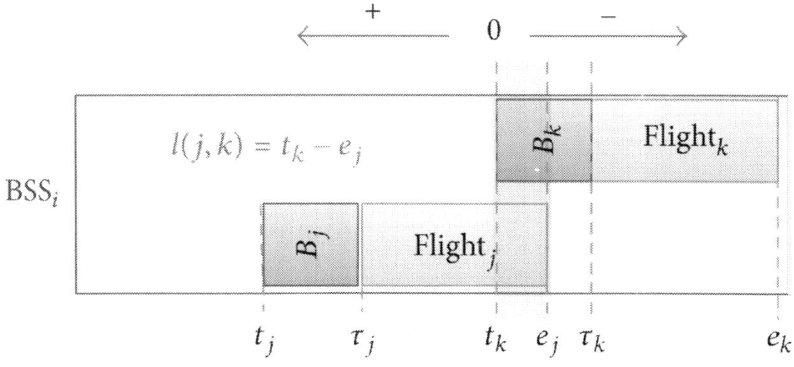

Figure 15: Overlap between two flights *j* and *k* assigned to the same BSS.

The definition of V(j, k) comes from the application domain, in the absence of historical data; some unsupervised estimation functions were introduced in [18]. Figure 16 shows the penalty (y-axis) incurred for different unsupervised estimation functions as a function of the gap (x-axis). Negative values refer to reductions in service time between two assignments to the same BSS, which are heavily penalized as they may require reassignment should delays occur, whereas positive gaps are penalized less. Wider gaps between two assignments reduce the need to reassign delayed flights, given that the delay has to be larger than the gap in order to affect the following assignment to the same BSS. Similarly, starting the service earlier may not require the flight to be reassigned because the duration of earliness has to be lower than the gap in order to affect the previous assignment. Both earliness and delay probabilities decrease as the gap increases. Sufficiently large gaps may also be used on the day of operation by disrupted flights which need to be reassigned, such that the detrimental effect of disruptions on that day is reduced.

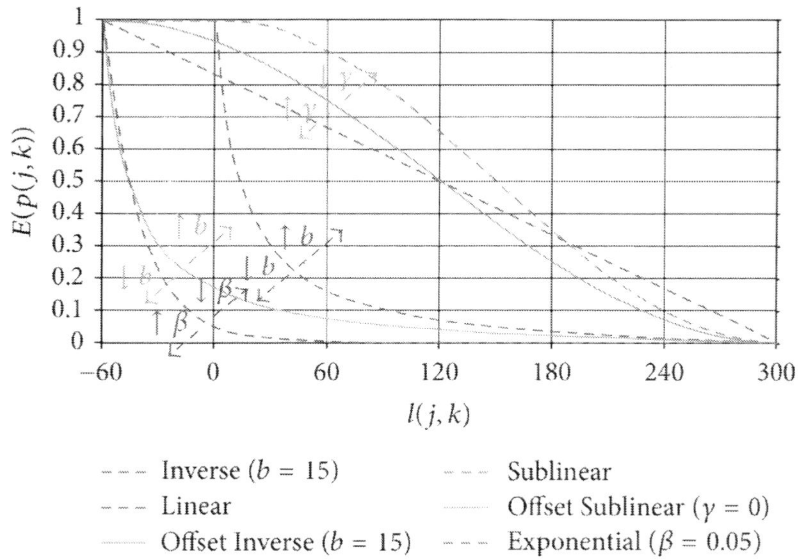

Figure 16: Penalty for different unsupervised estimation functions based on the gap between assignments.

The unsupervised estimation functions introduced in [18] are presented below and are shown in Figure 16:

(1) Linear estimation is as follows.

$$v(j,k) = -l(j,k). \tag{22}$$

(2) Exponential estimation is as follows.

$$v(j,k) = e^{-\beta * l(j,k)}. \tag{23}$$

(3) Inverse estimation is as follows.

$$v(j,k) = \begin{cases} \dfrac{b}{l(j,k)+b} & \text{if } l(j,k) > 0, \\ 1 & \text{otherwise.} \end{cases} \tag{24}$$

The value of the constant "*b*" changes the penalization as shown in Figure 16, so a higher "*b*" increases the penalization and a lower "*b*" decreases it. An appropriate value should be selected to properly weight the influence of the potential conflicts. Lim and Wang [18] used $b = 15$ minutes, which proved to provide rather poor results when compared with the exponential estimation function, which may partly be caused by the fixed cost when $l(j, k) < 0$ (dark red dash line, Figure 16), whereas in the exponential estimation function (purple dash line, Figure 16) this is not the case. The value used for "*b*" may be too great, and a lower value would make this estimation function provide values closer to those provided by the exponential estimation function which provided fitter solutions in the results presented in [18]. Consequently, a value b=6 was seen in the experiments studied in Section 4.1 to provide better results than when $b = 15$. In general, an even lower value did appear to perform better in some instances but not as well as b=6, as shown in Section 4.1.

The Inverse estimation function as considered in [18] treats all gaps smaller than the buffer time equally, which does not represent a real case since smaller gaps between flights are more likely to result in conflicts than larger ones on the day of operation. Given this and that the exponential estimation function performs best and treats all gaps differently, it is proposed that all of the gaps be treated differently, as shown by the modified version which is herein named "Offset inverse" (25), which is shown in Figure 16 for $b = 15$ (green line).

$$v(j,k) = \frac{b}{l(j,k) - \min\{l(j,k)\} + b}. \tag{25}$$

(4) Sublinear estimation is as follows.

$$v(j,k) = \begin{cases} \cos\left(\frac{\pi * l(j,k)}{l_{\max}}\right) & \text{if } l(j,k) > 0, \\ 1 & \text{otherwise.} \end{cases} \tag{26}$$

This estimation also suffers from the same problem as the Inverse estimation and may be improved by offsetting its value so that the maximum penalization corresponds to $l$max and the minimum corresponds to $l_{\min}$ (27), which is shown in Figure 16 for γ=0.

$$v(j,k) = \cos\left(\frac{\pi * (l(j,k) - l_{\min} + \gamma)}{l_{\max} - l_{\min} + \gamma}\right). \tag{27}$$

The gap definition used takes account of the buffer time, as the target service duration is the base service duration (T_k) plus the buffer time (B_k) for the flight. This makes the estimation functions dependent on the buffer time of each flight, as shown in Figure 17 for two buffer times of 30 and 15 min each. When the buffer time is the same irrespective of the flights, $B_k = B\ \forall k \in [1, \ldots, M]$, as considered in [18], the cost is the same irrespective of the flight, depending only on the separation between consecutive flight assignments, but this is not the situation when the buffer time depends on the flight, namely, long, medium, or short distance flights, which are the cases studied here.

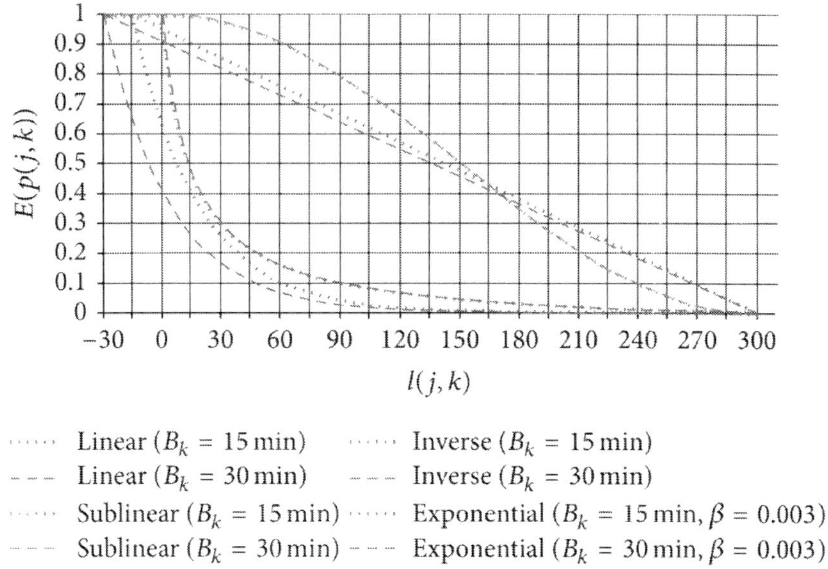

Figure 17: Penalty for different unsupervised estimation functions based on the gap between assignments for different buffer times.

Reduction in the Number of Conflicts (RNC)

A stochastic approach for improving schedule performance is described here, when disruptions occur on the day of operation. A similar approach was used in [19] where random delay scenarios are generated in the "planning stage" which are used to account for the potential disruptions in the schedule on the day of implementation by means of calculating the expected semideviation risk measure [20] for all those delay scenarios.

This approach is based on reducing the number of conflicts on the day of operation. Given that the real perturbed conditions will not be available until the day the schedule is implemented, these perturbed conditions are simulated by examining a set of perturbed base cases, S, which may be obtained in different ways, such as randomly, for example, from historical data or calculated using known distribution(s) from information available at the time of generating the assignments.

It is considered a set of perturbed schedules S, which simulate the perturbations on the day of operation. A new variable is introduced c_{js}, which for a given solution of assignments has the value of 1 if flight j is in conflict with another flight in the perturbed schedule $s \in S$ or zero otherwise. The average number of conflicts in the set of perturbed schedules S is calculated by (28), which is a measure of the solution robustness.

$$f_2 = \frac{-1}{|S|} * \sum_{s \in S} \sum_{j=1}^{M} c_{js}. \quad (28)$$

When reassigning conflicting flights or subsequent flights to other BSSs, a situation can arise where the reassigned flight interferes with the subsequent flight at the new BSS, a so-called secondary conflict. Some reassignment may therefore have a downstream effect on the overall schedule, producing more conflicting flights, in turn requiring further reassignments, thus potentially increasing the difficulty of the problem later on.

The above version considers all the conflicts to be of the same importance, but it is preferable to have conflicts which do not have repercussions later, that is, can be reassigned to another BSS without affecting any of the assignments already in existence. To account for this situation, a new variable c'_{js} is defined which takes the value of 1 if the reassignment of conflicting flight j in a perturbed schedule "s" affects other assignments already in existence or zero otherwise. The objective is presented as (29) where the constant, α, $0 \leq \alpha \leq 1$, denotes the importance of the conflicting flight repercussions on other assignments; $\alpha=0$ corresponds to the case where no account is taken of any repercussion on other assignments, which corresponds in turn to (28), and $\alpha=1$ corresponds to the cases in which both the conflicting flights and their repercussions on other assignments are considered to be of the same importance. $\alpha>1$ refers to the cases where more importance is given to the repercussions of a conflict on assignments other than the conflict itself.

$$f_2 = \frac{-1}{|S|} * \left(\sum_{s \in S} \sum_{j=1}^{M} \left(c_{js} + \alpha * c'_{js} \right) \right). \quad (29)$$

Calculation of the conflicts is time consuming and even more so if the effect of the conflict repercussions is also calculated, which is further aggravated by the need to use a large number of schedules in the perturbed set S in order to achieve a good representation of all the potential situations. The execution time is one of the disadvantages of using this approach as was experienced when executing the experiments presented in Section 4.

Probability of Conflict Based on the Gap (PCBG)

The previous approach would normally require a large number of perturbed data sets, which makes its application very slow. Given that we are still interested in reducing the number of conflicts, but without the heavy cost in speed imposed by the RNC approach, then it is proposed to use the probability of having a conflict in a given "idle time" for each flight. This can be easily obtained if the delay distribution is known. In the ABSSAP, the early arrival of a flight does not normally affect the assignment to the Baggage Sorting Station Selection (BSSS) as this does not extend the time in which the BSS is required for servicing the flight, but transportation of the baggage already in the BSS to the flight may start earlier, so no earlier arrival is considered here, in the case of a normal folded distribution (no negative numbers and with zero mean) and independent delays for flights. This could also be extended to other distributions and to nonzero means.

Independent delays are considered initially, where a conflict between two consecutive flights is independent of other flights assigned to the same BSS. The assignments to different BSSs are independent from the point of view of conflicts. The probability of two consecutive flights having overlapping service times (a conflict) corresponds to the sum of the product between the probability of a sufficiently large delay on the part of the previous flight assigned to that BSS and the probability of the next consecutive assignment to the same BSS not being sufficiently delayed, as shown in Figure 18, where $\eta(t)$ is the probability density function for flight j and $t_0 = \tau_k - e_j$.

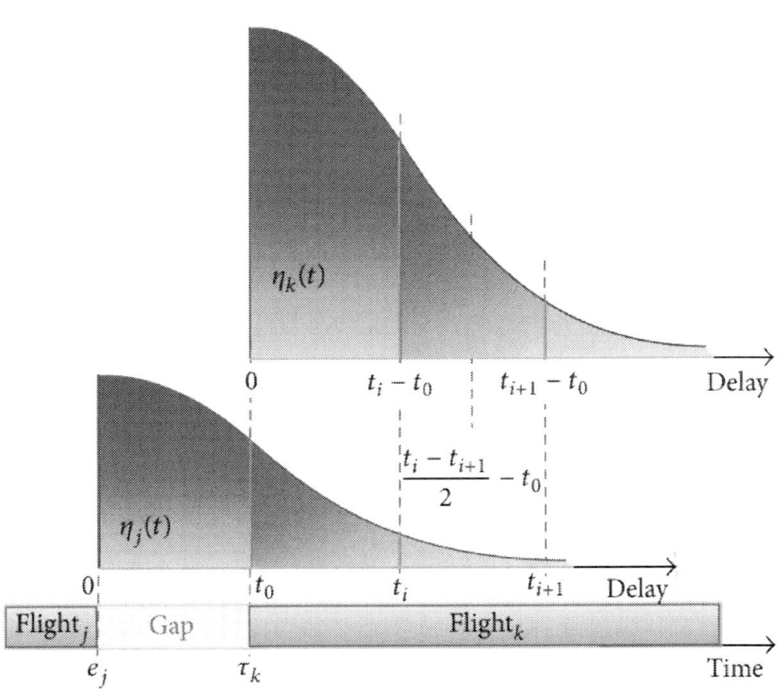

Figure 18: Probability of a conflict between two consecutive flights based on the intervening gap.

Equation (30) is the probability of a conflict, which is equal to the probability of exceeding the gap between both assignments, multiplied by the probability of the following assignment not being sufficiently delayed to avoid conflicts. $p(j, k)$is the probability of a conflict between two flights j and k assigned consecutively to the same BSS with a gap between them of $t_0 = \tau_k - e_j$ given their respective probability density distribution of delay $\eta_j(t)$ and $\eta_k(t)$.

$$p(j,k) = \int_{\tau_k - e_j}^{\infty} \eta_j(t) * \left(1 - \int_0^{t-(\tau_k - e_j)} \eta_k(x)\,dx\right) dt. \tag{30}$$

A simplification of the conflict probability is represented in (31), which uses the "Riemann integral" approach for the range of delays between $(\tau k - ej)$ and four times the delay distribution standard deviation of flight j, σ_j, and a time increment of Δt

$$p(j,k) = \sum_{i=0}^{(4*\sigma_j-(\tau_k-e_j))/\Delta t} \eta_j\left(\tau_k - e_j + \left(i + \frac{1}{2}\right) * \Delta t\right) * \left(1 - \sum_{l=0}^{i} \eta_k\left(\left(\left(l + \frac{1}{2}\right) * \Delta t\right) * \Delta t\right)\right) * \Delta t. \quad (31)$$

RESULTS

This section looks at the performance from the point of view of the robustness by measuring the number of conflicts for a given set of perturbed schedules, for all the approaches presented in Section 3: firstly when they are used alone, secondly when combined with the TRS approach, and finally for the combinations with the approaches which consider the flight densities. The comparison made between the results obtained, when applying the different approaches, uses the Mann-Whitney test to establish the statistical significance of the different approaches, presented in the summary result tables as the number of instances in each range of the number of BSSs which can be said to have no statistically significantly higher numbers of conflicts compared to any of the other approaches compared. Regarding an airport, *N* referring to the number of BSSs available, three ranges of the number of BSSs have been defined, based on the LMAP and UMAP, where the first is for $N <$ LMAP, the second is for LMAP $\leq N <$ UMAP, and the third is for UMAP $\leq N$. These are shown within brackets and separated by a comma in the following tables. Any approach achieving full coverage of a range of the number of BSSs is presented in bold font and the approaches with higher numbers for a range, covering the highest number of BSSs in the range compared, are presented in underlined font to assist in the interpretation of the summary result tables.

The robustness approaches described are applied to the ABSSAP using the Steady State Evolutionary Algorithm (SSEA) from [3] and their results are compared and analyzed in this section using the data sets obtained from NATS for London Heathrow airport Terminal 1, which were also used in [2, 3].

To compare the performance of each of the robustness approaches introduced in the previous sections, three sets of perturbed schedules were generated using a folded normal distribution with a zero mean and 10-, 20-, and 30-minute standard deviations. These sets are used to calculate the average number of conflicts for each robustness approach, where a lower value represents a more robust solution compared to those with higher values. Each experiment is repeated at least 30 times. To calculate the number of conflicts within a solution in the perturbed schedule, each flight in the original solution is assigned to the same BSS as in the original solution where possible; otherwise it is assigned to the dummy. This is repeated until all of the flights are assigned to BSS or the dummy. The number of flights assigned to the dummy represents the number of conflicts. If a solution does not achieve the maximum assignment possible, then the number of extra unassigned flights may be accounted for with a higher contribution, given that they are less desirable solutions. Nevertheless, all of the solutions having the highest fitness in the following experiments achieve maximum assignments, which simplifies comparison of the different robustness approaches, based on the described measure. This is possible since the measure only depends on assignments to the BSSs and not on any other objective, such as those upon which the fitness depends.

The number of possible perturbed schedules depends on the number of flights, and this accounts for an extremely large number of possible combinations, making it impossible to consider them all. The number of combinations for 194 flights corresponds to $194! \approx 1.3291 * 10^{361}$ which is far greater than 10^{82} atoms estimated to exist in the observable universe. The number of perturbed schedules necessary to calculate the quality of a solution should therefore be as large as possible to account for as many potential combinations as possible. However, as the number of perturbed schedules increases, so does the time required to perform the calculations, and the memory requirements also increase likewise: moreover, this is further multiplied by the number of solutions which will be used in the comparison. However, not all combinations are likely to represent a valid schedule. So, 10,000 perturbed schedules were used based on the number of solutions to be processed, the time available, and the memory required.

The stochastic robustness approach RNC is time consuming when compared with the other approaches reviewed. It would be desirable to use an approach which provides solutions closer to, or better than, those provided by the stochastic robustness approach, without the heavy cost of the time required. With the aim of assessing the difference in performance when the number of perturbed schedules is reduced, two sets of 1,000 and 25 perturbed schedules used with the RNC were also considered. Initial experiments were conducted using SSEA with l=1 and the RNC and Multiexchange between a Fixed Number of 3 Resources (MEFNR3) for 25 perturbed schedules and 800,000 total iterations which required an average execution time of 52 min per instance. These, when extrapolated to 1,000 perturbed schedules, provide an execution time of around 34 hours per instance, whereas the other approaches require no more than two minutes to complete the full set of iterations. Both applications of RNC required too long an execution time for the number of iterations specified, so they were run with a time limit of 30 min, as the other approaches required less than 2 min.

The objective importance presented in Section 2.1 is used in the following experiment where the most important objective is to achieve maximum assignment, with the second in importance being to maximize robustness and the third objective being minimization of the distance between flights and their assigned BSS being the last objective considered. The fitness function used to guide the search in the SSEA is a weighted sum of the different objectives evaluated, introduced previously in Section 2.1, whose weights are those calculated in [3]. Both approaches UESRS and PCBG need their robustness weight (W_2) to be recalculated. Thus, given a maximum distance between a flight and its assigned BSS, D_{max}, which depends on the airport topology, which for the topologies studied here is $D_{max} = 9$, a distance of one unit is assumed between different sides of a pier and a distance of two units was assumed between different piers. A new assignment between two previously assigned flights may incur a service reduction for the new assignment and next flight, which is used to obtain the decrease in robustness (the second objective), which for UESRS and PCBG cannot be greater than 1 for each flight, thus totaling 2 in this case. Finally, using the objective priorities the following relations can be established: $W_1 * 1 > |W_2| * 2 + |W_3| * D_{max}$ and $|W_2| * 2 > |W_3| * D_{max}$. The original conditions for the weights, when the "minimize reduction in service time" objective was used (Section 2.1), are $W3 = -1$ and $W_1 >$ 23.4, which together with the objective priorities obtain $|W_2| > (1 * 9)/2 =$ 4.5 and $|W_2|*2+|W_3|*9 < 23.4$ giving $|W_2| < 14.4/2 = 7.2$.The value used for W_2 is −7.2 as $W_1 = 90 > 23.4$. The value used for W_1 is greater than the value originally used to calculate W_2, so a value of −10 was also used.

The unsupervised estimation functions introduced in Section 3.6 were used for the same parameter values as those used in [18] and with Offset Inverse b=6, Offset Inverse b = 15, Offset Sublinear γ=0, and Offset Sublinear γ = 1000. A summary of the robustness approaches studied and their parameter values is shown in Table 3. In Table 3, the first column contains the name of the robustness approaches considered, all of which were introduced in Section 3; the second column shows the weights for each approach, and the subsequent columns show the name and value of the parameters for the corresponding robustness approach. The SSEA with ℓ=1, operator MEFNR3 from [2, 3], a population size of 10, the replacement strategy being Index Selection with Elitist Selection and a group size of 1 (IS1ES), and a maximum of 800,000 iterations for two data sets from London Heathrow airport Terminal 1 (H1T091216: 16th December 2009 and H1T100301: 1st March 2010) were used to obtain the solutions for comparison. The fitness is the sum of the total number of assignments with weight of 90, less the distance between assigned flights and their assigned BSS with a weight of 1, less the robustness approach with the appropriate weight, all of which are shown in Table 3.

Table 3: Robustness approaches used with their parameter values

Approach	**Weight (W_2)**	**Parameters Name: values**
TRS	0.008	Buffer time: 30 min. long-haul and 15 min. others
ATRS	0.008	Buffer time: 30 min. long-haul and 15 min. others
ARS	0.008	Buffer time: 30 min. long-haul and 15 min. others
BSARS	0.008	Buffer time: 30 min. long-haul and 15 min. others
SARS	0.008	Buffer time: 30 min. long-haul and 15 min. others
PCBG	7.2 and 10	Std. deviation: 10, 20 and 30 min.
RNC	10 and 14	Std. deviation: 10, 20 and 30 min.
		Num. schedules: 25 and 1000
		Max. execution time: 30 min.

UESRS	7.2 and 10	Estimation function: Exp 0.03, Exp 0.05, Inverse 6, Inverse 15, Linear, Offset Inverse 6, Offset Inverse 15, Offset Sublinear 0, Offset Sublinear 1000, and Sublinear
		Buffer time: 30 min. long-haul and 15 min. others

The results presented in the following sections were summarized for simplicity and clarity by considering the average number of times an approach achieves statistically significantly lower conflicts, or at least no worse, than the other approaches, in the different regions of numbers of BSSs, based on the LMAP and UMAP, which divide the range of BSSs studied into three areas corresponding to $N <$ LMAP, LMAP $\leq N <$ UMAP, and UMAP $\leq N$. The values between brackets correspond to the number of times the approach provides significantly statistically solutions not worse than the other approaches used, for each of the BSS ranges. Table 4 shows that the ARS approach for the data set of H1T091216 as well as a 4-pier topology has the values (2, 3, 3) showing that it achieves a statistically significant number of conflicts not higher in two instances for the range of $N <$ LMAP, three for the range LMAP $\leq N <$ UMAP, and three for $N \geq$ UMAP. Thus, the larger the number between parenthesis, the better the performances in respect of robustness.

Table 4: Number of instances with significantly statistically lower number of conflicts in each range of numbers of BSSs for disruptions with a delay = 10 min for ARS, BSARS, SARS + TRS, and TRS for a significance level of 0.05

Approach	H1T091216 Max. (9, 5, 3)		H1T100301 Max. (6, 6, 5)	
	3-pier	4-pier	3-pier	4-pier
ARS	(1, 2, 2)	(2, 3, **3**)	(3, 3, 4)	(1, 2, 1)
BSARS	(**9**, **5**, **3**)	(8, 2, **3**)	(4, **6**, **5**)	(4, 5, **5**)
SARS + TRS	(3, 4, 2)	(3, **5**, **3**)	(3, 3, 4)	(0, 2, 1)
TRS	(0, 1, 1)	(1, 0, 0)	(2, 3, 4)	(1, 2, 0)

Bold font is used to identify those cases where the robustness approach achieves good results for all numbers of BSSs in a range; for example, in the previous example the approach performs well for all numbers of BSSs in the range of $N \geq 27$ (UMAP). Underlining is used to identify those cases where the robustness approach performs well for greater numbers of BSSs in a range. The maximum quantity of instances of number of BSSs in a range is presented between brackets at the top of the table for each of the ranges discussed, preceded by the word "Max."; for example, in Table 4 the cell in the second column and second row of the header shows that the first range contains nine instances of numbers of BSSs for $N <$ LMAP, five for LMAP $\leq N <$ UMAP, and three for $N \geq$ UMAP. For simplicity and clarity, those approaches which do achieve statistically significantly higher number of conflicts on average than any of the other approaches in all three BSSs regions are not shown in the tables which followed.

The next sections present the experimental results for the robustness approaches presented.

Results for All Robustness Approaches Considered

In this section, we look at the performance from the point of view of robustness for all the approaches presented in Section 3; first when they are used alone and then when combined with the TRS approach.

The BSARS provides more robust solutions than those obtained by the other proposed approaches, that is, TRS, ARS, SARS, and TRS + SARS, as shown in Table 4. Furthermore, BSARS will be seen to perform even better when combined with TRS in Tables 5, 6, and 7. ARS and SARS were similarly seen to improve robustness when combined with TRS.

Table 5: Number of instances with a significantly statistically lower number of conflicts in each range of numbers of BSSs for disruptions with a delay σ = 10 min and all the approaches alone and combined with TRS for a significance level of 0.05

Approach	**H1T091216**		**H1T100301**	
	Max. (9, 5, 3)		Max. (6, 6, 5)	
	3-pier	4-pier	3-pier	4-pier
BSARS	(2, 0, 0)	(1, 0, 0)	(0, 0, 0)	(3, 0, 0)
TRS + BSARS	(3, 0, 0)	(3, 0, 0)	(0, 0, 0)	(4, 0, 0)
PCBG				
7.2	(5, 2, 1)	(3, 1, 0)	(1, 1, 0)	(5, 2, 0)
10	(7, 4, 0)	(8, 1, 0)	(2, 4, 0)	(5,4 , 0)

TRS + UESRS				
Exp 0.03				
7.2	(2, 1, 0)	(0, 1, 0)	(2, 1, 1)	(1, 2, 0)
10	(2, 1, 0)	(1, 4, 3)	(4, 2, 4)	(2, 1, 1)
Inverse 6				
7.2	(0, 0,2)	(0, 1, 0)	(0, 1, 1)	(0, 0, 0)
10	(0, 0, 2)	(0, 1, 2)	(0, 1, 4)	(0, 2, 4)

Table 6: Number of instances with a significantly statistically lower number of conflicts in each range of numbers of BSSs for disruptions with a delay σ = 20 min and all the approaches alone and combined with TRS for a significance level of 0.05

Approach	**H1T091216**		**H1T1003010**	
	Max. (9, 5, 3)		Max. (6, 6, 5)	
	3-pier	4-pier	3-pier	4-pier
BSARS	(0, 0, 0)	(1, 0, 0)	(0, 0, 0)	(2, 0, 0)
TRS + BSARS	(0, 0, 0)	(4, 0, 0)	(0, 0, 0)	(3, 0, 0)
PCBG				
7.2	(5, 1, 0)	(2, 0, 0)	(3, 2, 0)	(3, 0, 0)
10	(7, 2, 1)	(8, 5, 1)	(1, 5, 1)	(5, 6, 1)
TRS + UESRS				
Exp 0.03				
7.2	(2, 1, 0)	(1, 0, 0)	(2, 1, 0)	(1, 0, 0)
10	(5, 3, 1)	(2, 0, 3)	(5, 1, 4)	(4, 0, 1)
Inverse 6				
10	(1, 1, 3)	(0, 2, 2)	(0, 1, 2)	(2, 2, 5)

Table 7: Number of instances with a significantly statistically lower number of conflicts in each range of numbers of BSSs for disruptions with a delay σ = 30 min and all the approaches alone and combined with TRS for a significance level of 0.05

Approach	**H1T091216**		**H1T100301**	
	Max. (9, 5, 3)		Max. (6, 6, 5)	
	3-pier	4-pier	3-pier	4-pier
BSARS	(0, 0, 0)	(1, 0, 0)	(0, 0, 0)	(1, 0, 0)
TRS + BSARS	(0, 0, 0)	(2, 0, 0)	(0, 0, 0)	(2, 0, 0)
PCBG				
7.2	(4, 2, 0)	(0, 0, 0)	(1, 4, 0)	(1, 0, 0)
10	(1, 4, 3)	(5, 4, 3)	(2, 6, 5)	(3, 6, 5)
TRS + UESRS				
Exp 0.03				
7.2	(3, 0, 0)	(1, 0, 0)	(3, 0, 0)	(1, 0, 0)
10	(9, 1, 0)	(4, 1, 0)	(5, 1, 0)	(2, 0, 0)
Inverse 6				
7.2	(1, 0, 0)	(1, 0, 0)	(1, 0, 0)	(2, 0, 0)
10	(3, 3, 1)	(2, 3, 0)	(0, 0, 0)	(3, 0, 0)

The results which are summarized in Tables 5, 6, and 7 show that the PCBG does not gain any advantage when combined with the TRS, but it appears to be detrimental, as PCBG alone performs better throughout the whole ranges of the numbers of BSSs compared to when used combined with TRS. Even with the RNC for 25 and 1,000 instances taking account of the distribution of delays, the other approaches provide a statistically significantly lower number of collisions, but with a much lower running time. This could be due to the fact that the number of iterations which it is possible to execute in the 30 minutes is too low to find promising solutions with a lower number of collisions compared to those obtained by the other approaches. Around 33,000 to 70,000 iterations were executed for the experiments conducted here (the numbers depending mainly on the data set and the number of BSSs).

The PCBG provides statistically significantly lower number of conflicts through a wider range of BSSs, and such a range also includes the range of BSSs used in the real problems; that is, $N \geq$ UMAP.

Nevertheless, this result could be regarded as biased, given that the PCBG considers a normal folded distribution of the same standard deviation as that from which the perturbed schedules were generated. The PCBG could consider different standard deviations and distributions depending on the aircraft, season, route, destination, and time of the day which should further improve the results in real situations. The running time is also of the same magnitude as that for the other approaches, with the exception of the RNC, which has a much higher running time for the same number of iterations.

There is a possibility of combining the different approaches, in particular either ARS or BSARS with UESRS studied below. There is also a question about the preference as to how these approaches are combined, either as a sum of each individual with the TRS, as studied in here, or as a product of their individual contributions.

Other Combinations of These Approaches

In this section we study the combination of the UESRS with each of the ARS, BSARS, and TRS. It may be noted that the SARS is equal to BSARS divided by (UMAP-LMAP), which it is equivalent to thus changing the weight (W_2) in the fitness function. As seen previously in Table 4, BSARS provides more robust solutions compared to when using SARS alone for the same weight (W_2) or when combined with TRS, and these differences between both similar approaches are even more pronounced when BSARS is used in combination with TRS.

The combination of TRS and UESRS with exponential estimation function overall performs better for $N <$ LMAP, as shown in Tables 8, 9, and 10, which correspond to the situation when there are no sufficient resources for the number of flights to service. The results which are summarized in Tables 11, 12, and 13 and which correspond to the cases when two robustness approaches are combined show that the combination of ARS with UESRS for exponential estimation function overall performs better for LMAP ≤N< UMAP, which refers to the situation when there are just sufficient resources for the number of flights to service, as, in this case, it is particularly important to consider the distribution of the flights requiring service which ARS takes into account. Also the combination of UESRS with BSARS in general improves the performance more than when using ARS for all the ranges. Similarly, the combination of BSARS with UESRS with Inverse estimation function overall performs better for UMAP ≤ N and also for N lower than UMAP but closer to it. This would be the preferable approach for real problems. Finally, the ARS with UESRS with Inverse estimation function performance decreases for big delays for N ≥ UMAP and improves for N ≤ LMAP.

Table 8: Terminal 1 with algorithm SSEA1, operator MEFNR3, and population size of 2,000 for perturbations of 0 min average and σ = 10 min deviation, Mann-Whitney test, and significance level of 0.05 and for the TRS, UESRS, and the combination of TRS with UESRS

Approach	H091216		H100301	
	17 (9, 5, 3)		17 (6, 6, 5)	
	3-pier	4-pier	3-pier	4-pier
0.008 × TRS + 10 × UESRS (Exp 0.03)	14 (**9**, 4, 1)	15 (7, **5**, **3**)	14 (**6**, 5, 3)	11 (**6**, 4, 1)
0.008 × TRS + 10 × UESRS (Inverse 2)	1 (0, 0, 1)	2 (0, 2, 0)	1 (0, 0, 1)	0 (0, 0, 0)
0.008 × TRS + 10 × UESRS (Inverse 4)	2 (0, 0, 2)	3 (0, 2, 1)	3 (0, 1, 2)	6 (0, 2, 4)
0.008 × TRS + 10 × UESRS (Inverse 6)	3 (0, 1, 2)	4 (0, 2, 2)	4 (0, 1, 3)	6 (0, 2, 4)
0.008 × TRS + 7.2 × UESRS (Exp 0.03)	8 (7, 1, 0)	5 (4, 1, 0)	9 (4, 4, 1)	8 (4, 4, 0)
0.008 × TRS + 7.2 × UESRS (Inverse 2)	1 (0, 0, 1)	1 (0, 1, 0)	0 (0, 0, 0)	0 (0, 0, 0)
0.008 × TRS + 7.2 × UESRS (Inverse 4)	1 (0, 0, 1)	1 (0, 1, 0)	2 (0, 1, 1)	1 (0, 1, 0)
0.008 × TRS + 7.2 × UESRS (Inverse 6)	2 (0, 0, 2)	1 (0, 1, 0)	2 (0, 1, 1)	0 (0, 0, 0)
10 × UESRS (Exp 0.03)	0 (0, 0, 0)	1 (1, 0, 0)	2 (0, 0, 2)	0 (0, 0, 0)

Table 9: Terminal 1 with algorithm SSEA1, operator MEFNR3, and population size of 2,000 for perturbations of 0 min average and σ = 20 min deviation, Mann-Whitney test, and significance level of 0.05 and for the TRS, UESRS, and the combination of TRS with UESRS

Approach	H091216		H100301	
	Max. 17 (9, 5, 3)		Max. 17 (6, 6, 5)	
	3-pier	4-pier	3-pier	4-pier
0.008 × TRS + 10 × UESRS (Exp 0.03)	12 (**9**, 3, 0)	13 (7, 3, **3**)	14 (**6**, 4, 4)	9 (5, 3, 1)
0.008 × TRS + 10 × UESRS (Inverse 2)	2 (2, 0, 0)	4 (4, 0, 0)	0 (0, 0, 0)	2 (2, 0, 0)
0.008 × TRS + 10 × UESRS (Inverse 4)	5 (2, 1, 2)	5 (2, 3, 0)	1 (0, 0, 1)	5 (2, 1, 2)
0.008 × TRS + 10 × UESRS (Inverse 6)	5 (0, 2, **3**)	10 (3, **5**, 2)	4 (0, 1, 3)	12 (3, **4**, **5**)
0.008 × TRS + 7.2 × UESRS (Exp 0.03)	4 (3, 1, 0)	3 (3, 0, 0)	4 (3, 1, 0)	4 (2, 2, 0)
0.008 × TRS + 7.2 × UESRS (Inverse 2)	1 (0, 1, 0)	1 (1, 0, 0)	0 (0, 0, 0)	1 (1, 0, 0)
0.008 × TRS + 7.2 × UESRS (Inverse 4)	1 (1, 0, 0)	1 (1, 0, 0)	1 (0, 1, 0)	0 (0, 0, 0)
0.008 × TRS + 7.2 × UESRS (Inverse 6)	0 (0, 0, 0)	0 (0, 0, 0)	1 (0, 1, 0)	1 (0, 1, 0)
10 × UESRS (Exp 0.03)	0 (0, 0, 0)	1 (1, 0, 0)	1 (0, 0, 1)	0 (0, 0, 0)

Table 10: Terminal 1 with algorithm SSEA1, operator MEFNR3, and population size of 2,000 for perturbations of 0 min average and σ = 30 min deviation, Mann-Whitney test, and significance level of 0.05 and for the TRS, UESRS, and the combination of TRS with UESRS

Approach	H091216		H100301	
	Max. 17 (9, 5, 3)		Max. 17 (6, 6, 5)	
	3-pier	4-pier	3-pier	4-pier
0.008 × TRS + 10 × UESRS (Exp 0.03)	13 (**9**, 3, 1)	12 (6, 3, **3**)	15 (**6**, 5, 4)	6 (3, 2, 1)
0.008 × TRS + 10 × UESRS (Inverse 2)	5 (5, 0, 0)	3 (3, 0, 0)	0 (0, 0, 0)	4 (4, 0, 0)
0.008 × TRS + 10 × UESRS (Inverse 4)	7 (5, 1, 1)	8 (5, 3, 0)	3 (1, 1, 1)	3 (3, 0, 0)
0.008 × TRS + 10 × UESRS (Inverse 6)	9 (3, 3, **3**)	11 (5, **5**, 1)	10 (2, 5, 3)	12 (3, 4, **5**)
0.008 × TRS + 7.2 × UESRS (Exp 0.03)	3 (3, 0, 0)	2 (2, 0, 0)	3 (3, 0, 0)	2 (1, 1, 0)
0.008 × TRS + 7.2 × UESRS (Inverse 2)	1 (1, 0, 0)	3 (3, 0, 0)	0 (0, 0, 0)	2 (2, 0, 0)
0.008 × TRS + 7.2 × UESRS (Inverse 4)	3 (3, 0, 0)	2 (2, 0, 0)	1 (0, 1, 0)	1 (1, 0, 0)
0.008 × TRS + 7.2 × UESRS (Inverse 6)	1 (1, 0, 0)	2 (2, 0, 0)	3 (2, 1, 0)	2 (2, 0, 0)
10 × UESRS (Exp 0.03)	0 (0, 0, 0)	0 (0, 0, 0)	2 (1, 0, 1)	0 (0, 0, 0)

Table 11: Conflicts statistical significance for combined robustness approaches for Terminal 1 with algorithm SSEA1, operator MEFNR3, and population size of 2,000 for perturbations of 0 min average and σ = 10 min deviation, Mann-Whitney test, and significance level of 0.05

Approach	H091216		H100301	
	Max. 17 (9, 5, 3)		Max. 17 (6, 6, 5)	
	3-pier	4-pier	3-pier	4-pier
0.008 × ARS + 10 × UESRS (Exp 0.03)	3 (1, 2, 0)	5 (0, 3, 2)	4 (0, 1, 3)	3 (0, 2, 1)
0.008 × ARS + 10 × UESRS (Inverse 6)	1 (0, 0, 1)	4 (0, 1, **3**)	3 (0, 1, 2)	6 (0, 1, **5**)
0.008 × ARS + 7.2 × UESRS (Inverse 6)	3 (0, 2, 1)	0 (0, 0, 0)	1 (0, 0, 1)	0 (0, 0, 0)
0.008 × BSARS + 10 × UESRS (Exp 0.03)	7 (4, 2, 1)	6 (0, 3, 3)	3 (0, 0, 3)	6 (3, 2, 1)
0.008 × BSARS + 10 × UESRS (Inverse 6)	7 (1, 3, 3)	5 (0, 2, 3)	6 (0, 2, 4)	6 (0, 2, 4)
0.008 × BSARS + 7.2 × UESRS (Exp 0.03)	3 (2, 1, 0)	2 (2, 0, 0)	1 (0, 0, 1)	5 (4, 1, 0)
0.008 × BSARS + 7.2 × UESRS (Inverse 6)	5 (1, 2, 2)	0 (0, 0, 0)	2 (0, 1, 1)	0 (0, 0, 0)
0.008 × TRS + 0.008 × BSARS	0 (0, 0, 0)	2 (2, 0, 0)	0 (0, 0, 0)	1 (1, 0, 0)
0.008 × TRS + 10 × UESRS (Exp 0.03)	0 (0, 0, 0)	2 (0, 0, 2)	4 (2, 0, 2)	1 (0, 0, 1)
0.008 × TRS + 10 × UESRS (Inverse 2)	1 (0, 0, 1)	0 (0, 0, 0)	1 (0, 0, 1)	0 (0, 0, 0)
0.008 × TRS + 10 × UESRS (Inverse 4)	1 (0, 0, 1)	0 (0, 0, 0)	2 (0, 0, 2)	4 (0, 1, 3)
0.008 × TRS + 10 × UESRS (Inverse 6)	2 (0, 0, 2)	1 (0, 0, 1)	3 (0, 0, 3)	5 (0, 1, 4)
0.008 × TRS + 7.2 × PCBG (10)	6 (6, 0, 0)	2 (2, 0, 0)	4 (2, 2, 0)	4 (2, 2, 0)
0.008 × TRS + 7.2 × UESRS (Inverse 4)	1 (0, 0, 1)	0 (0, 0, 0)	1 (0, 0, 1)	0 (0, 0, 0)
0.008 × TRS + 7.2 × UESRS (Inverse 6)	1 (0, 0, 1)	0 (0, 0, 0)	1 (0, 0, 1)	0 (0, 0, 0)
10 × PCBG (10)	5 (4, 1, 0)	5 (5, 0, 0)	6 (2, 4, 0)	5 (2, 3, 0)
7.2 × PCBG (10)	0 (0, 0, 0)	0 (0, 0, 0)	1 (0, 1, 0)	3 (1, 2, 0)

Table 12: Conflicts statistical significance for combined robustness approaches for Terminal 1 with algorithm SSEA1, operator MEFNR3, and population size of 2,000 for perturbations of 0 min average and σ = 20 min deviation, Mann-Whitney test, and significance level of 0.05

Approach	H091216		H100301	
	Max. 17 (9, 5, 3)		Max. 17 (6, 6, 5)	
	3-pier	4-pier	3-pier	4-pier
0.008 × ARS + 10 × UESRS (Exp 0.03)	3 (1, 2, 0)	0 (0, 0, 0)	3 (0, 1, 2)	1 (0, 0, 1)
0.008 × ARS + 10 × UESRS (Inverse 6)	3 (0, 1, 2)	0 (0, 0, 0)	3 (0, 0, 3)	7 (0, 2, **5**)
0.008 × BSARS + 10 × UESRS (Exp 0.03)	7 (3, 2, 2)	0 (0, 0, 0)	5 (0, 1, 4)	3 (3, 0, 0)
0.008 × BSARS + 10 × UESRS (Inverse 6)	10 (3, 4, **3**)	0 (0, 0, 0)	6 (0, 3, 3)	9 (2, 3, 4)
0.008 × BSARS + 7.2 × UESRS (Exp 0.03)	3 (2, 0, 1)	0 (0, 0, 0)	0 (0, 0, 0)	3 (3, 0, 0)
0.008 × BSARS + 7.2 × UESRS (Inverse 6)	2 (2, 0, 0)	0 (0, 0, 0)	0 (0, 0, 0)	1 (1, 0, 0)
0.008 × TRS + 0.008 × ARS	0 (0, 0, 0)	**17** (**9**, **5**, **3**)	0 (0, 0, 0)	0 (0, 0, 0)
0.008 × TRS + 10 × UESRS (Exp 0.03)	0 (0, 0, 0)	0 (0, 0, 0)	2 (1, 0, 1)	1 (0, 0, 1)
0.008 × TRS + 10 × UESRS (Inverse 6)	1 (0, 0, 1)	0 (0, 0, 0)	2 (0, 0, 2)	5 (0, 1, 4)
0.008 × TRS + 7.2 × PCBG (20)	7 (7, 0, 0)	0 (0, 0, 0)	8 (4, 4, 0)	3 (2, 1, 0)
10 × PCBG (20)	4 (3, 1, 0)	0 (0, 0, 0)	6 (1, 4, 1)	9 (2, **6**, 1)

Table 13: Conflicts statistical significance for combined robustness approaches for Terminal 1 with algorithm SSEA1, operator MEFNR3, and population size of 2,000 for perturbations of 0 min average and σ = 30 min deviation, Mann-Whitney test, and significance level of 0.05

Approach	H091216		H100301	
	Max. 17 (9, 5, 3)		Max. 17 (6, 6, 5)	
	3-pier	4-pier	3-pier	4-pier
0.008 × ARS + 10 × UESRS (Exp 0.03)	3 (2, 1, 0)	1 (0, 1, 0)	3 (2, 1, 0)	0 (0, 0, 0)
0.008 × ARS + 10 × UESRS (Inverse 6)	2 (0, 1, 1)	1 (0, 1, 0)	0 (0, 0, 0)	1 (1, 0, 0)
0.008 × ARS + 7.2 × UESRS (Exp 0.03)	0 (0, 0, 0)	1 (1, 0, 0)	1 (1, 0, 0)	0 (0, 0, 0)
0.008 × ARS + 7.2 × UESRS (Inverse 6)	0 (0, 0, 0)	0 (0, 0, 0)	1 (1, 0, 0)	1 (1, 0, 0)
0.008 × BSARS + 10 × UESRS (Exp 0.03)	6 (3, 2, 1)	0 (0, 0, 0)	4 (2, 1, 1)	4 (3, 1, 0)
0.008 × BSARS + 10 × UESRS (Inverse 6)	11 (6, 4, 1)	1 (0, 1, 0)	3 (1, 2, 0)	5 (3, 2, 0)
0.008 × BSARS + 7.2 × UESRS (Exp 0.03)	2 (2, 0, 0)	4 (4, 0, 0)	1 (1, 0, 0)	3 (3, 0, 0)
0.008 × BSARS + 7.2 × UESRS (Inverse 6)	2 (2, 0, 0)	5 (5, 0, 0)	1 (1, 0, 0)	2 (2, 0, 0)

0.008 × TRS + 10 × UESRS (Exp 0.03)	1 (1, 0, 0)	1 (1, 0, 0)	3 (3, 0, 0)	1 (1, 0, 0)
0.008 × TRS + 10 × UESRS (Inverse 6)	3 (1, 1, 1)	1 (1, 0, 0)	0 (0, 0, 0)	0 (0, 0, 0)
0.008 × TRS + 7.2 × PCBG (30)	13 (7, **5**, 1)	5 (3, 2, 0)	11 (5, **6**, 0)	5 (2, 3, 0)
0.008 × TRS + 7.2 × UESRS (Exp 0.03)	2 (2, 0, 0)	1 (1, 0, 0)	0 (0, 0, 0)	0 (0, 0, 0)
10 × PCBG (30)	5 (0, 2, **3**)	11 (4, 4, **3**)	9 (2, 2, **5**)	13 (2, **6**, **5**)

Future studies could look at the dependency of the robustness objective weight on performance. If other robustness weights are to be considered, all of the weights taking part in the fitness function should be modified accordingly to maintain the order of importance of each objective. When the unassigned flights are not taken into account by the robustness measure used in the fitness function, then particular care has to be taken when selecting the appropriate weights for maximization of the number of assignments (the first objective), since incorrect selection of this objective weight may sometimes deem solutions with a lower number of assignments to be fitter. This interdependency mainly between the maximum number of assignments' objective and the robustness could also be decoupled by penalizing the unassigned flights in the robustness.

CONCLUSIONS

Several approaches were presented here to take account of solution robustness when applying the SSEA. The TRS approach provides solutions with a statistically significantly higher number of conflicts than those obtained by many of the other approaches considered, as shown in [21]. TRS does not consider the extra increase in conflict, as the service time is reduced between assignments, thus further penalizing those assignments. Where the service reduction is higher the Arc Tangent Reduction in Service (ATRS) was also used, but the results were not very encouraging as some of the other approaches improved on it.

When looking at the overall schedule it is evident that the number of flights requiring assignment at each time is not uniformly distributed, as shown in Figure 2, such that when the number of flights to be serviced at any one time increases, the potential for conflict also increases, when perturbed, such that there is potential for further improving the results when this is taken into account, so other approaches were proposed and studied. The approaches examined, which consider the distribution of flights in time, were the ARS, BSARS, and SARS. These approaches performed better for a very low number of BSSs but in general provided solutions inferior in respect of conflicts compared to the PCBG, as well as the UESRS with exponential and Inverse estimation functions. The advantage of the TRS approach in conjunction with some of the other approaches was anticipated, as was corroborated by results from the experiments conducted for the combination of TRS with the ARS, BSARS, SARS, and UESRS.

Several stochastic approaches were also considered; namely, UESRS was evaluated for different estimation functions and the exponential function with $\beta = 0.03$ provided the best results in respect of robustness for the measure considered. PCBG provides the best overall performance regarding robustness but did not seem to improve when combined with TRS. RNC appeared to provide solutions with a higher number of conflicts compared to the other approaches, but this could be a consequence of the low number of iterations which it was possible to execute given both the low speed, which is a characteristic of this approach, and the higher memory requirement. The UESRS approach also improved when combined with TRS, showing the Inverse function for b=6 with TRS providing good results for high numbers of BSSs ($N \geq$ LMAP).

When UESRS was combined with BSARS the performance improved to when these approaches where used alone or combined with TRS.

It would be interesting to apply these approaches to some data sets where the original flight schedules and final real schedules are known, to see if the solutions obtained by these approaches did cope well with the changes, but unfortunately this was impossible at the time, given the unavailability of such data. Furthermore, if the data sets contain the real flight assignments to BSSs on the day of implementation, then it would be possible to quantify the actual improvement which could have been achieved by each of the approaches presented here, if they were implemented. There is also a question about the preference as to how these approaches are combined, either as a sum of each individual with the TRS, BSARS, and ARS, as studied here, or as a product of their individual contributions.

Future work should consider the use of multiple distributions, in general one per flight, based on the particular characteristics of each flight, such as aircraft type, airline, destination, route, and season. This could be applied similarly to the generation of the buffer times. This approach was not used, since such information was unavailable at the time this study was conducted. It is envisaged that the use of this information in the PCBG and RNC will improve their performance and may also be used to generate the disrupted schedules used to measure the solution quality provided by all of the approaches, thus assisting in the identification of the best approach for use in the specific problem.

More information about the problem studied here, the metaheuristic algorithms used, and the application of these approaches to the ABSSAP and AGAP can be found in [21].

ACKNOWLEDGMENTS

The author is grateful to NATS Ltd. and to EPSRC for providing the funding for the original project and to the University of Nottingham for giving the author the opportunity to take part in this project.

REFERENCES

1. J. M. Mulvey, R. J. Vanderbei, and S. A. Zenios, "Robust optimization of large-scale systems," Operations Research, vol. 43, no. 2, pp. 264–281, 1995.
2. A. Ascó, J. A. D. Atkin, and E. K. Burke, "An analysis of constructive algorithms for the airport baggage sorting station assignment problem," Journal of Scheduling, vol. 17, no. 6, pp. 601–619, 2014.
3. A. Ascó, J. A. D. Atkin, and E. K. Burke, "An evolutionary algorithm for the over-constrained airport baggage sorting station assignment problem," in Simulated Evolution and Learning: 9th International Conference, SEAL 2012, Hanoi, Vietnam, December 16–19, 2012. Proceedings, L. Bui, Y. Ong, N. Hoai, H. Ishibuchi, and P. Suganthan, Eds., vol. 7673 of Lecture Notes in Computer Science, pp. 32–41, Springer, Berlin, Germany, 2012.
4. A. Ascó, J. A. D. Atkin, and E. K. Burke, "The airport baggage sorting station allocation problem," in Proceedings of the 5th Multidisci-Plinary International Conference on Scheduling: Theory and Applications (MISTA ‹11), J. Fowler, G. Kendall, and B.

McCollum, Eds., pp. 419–444, Phoenix, Ariz, USA, August 2011.

5. A. Abdelghany, K. Abdelghany, and R. Narasimhan, "Scheduling baggage-handling facilities in congested airports," Journal of Air Transport Management, vol. 12, no. 2, pp. 76–81, 2006.
6. Y. Nikulin, "Robustness in combinatorial optimization and scheduling theory: an extended annotated bibliography," Tech. Rep. 606, Institut für Betriebswirtschaftslehre, 2006.
7. C.-L. Wu and R. E. Caves, "Modelling and optimization of aircraft turnaround time at an airport," Transportation Planning and Technology, vol. 27, no. 1, pp. 47–66, 2004.
8. S. Yan, C.-Y. Shieh, and M. Chen, "A simulation framework for evaluating airport gate assignments," Transportation Research Part A: Policy and Practice, vol. 36, no. 10, pp. 885–898, 2002.·
9. A. Bolat, "Procedures for providing robust gate assignments for arriving aircrafts," European Journal of Operational Research, vol. 120, no. 1, pp. 63–80, 2000.
10. S. Yan and C. M. Chang, "A network model for gate assignment," Journal of Advanced Transportation, vol. 32, no. 2, pp. 176–189, 1998.
11. M. I. Hassounah and G. N. Steuart, "Demand for aircraft gates," Transportation Research Record, no. 1423, pp. 26–33, 1993.
12. S. Yan and C.-M. Huo, "Optimization of multiple objective gate assignments," Transportation Research Part A: Policy and Practice, vol. 35, no. 5, pp. 413–432, 2001.
13. D.-X. Wei and C.-Y. Liu, "Fuzzy model and optimization for airport gate assignment problem," in Proceedings of the IEEE International Conference on Intelligent Computing and Intelligent Systems (ICIS ‹09), pp. 828–832, Shanghai, China, November 2009.
14. C.-L. Wu and R. E. Caves, "Aircraft operational costs and turnaround efficiency at airports," Journal of Air Transport Management, vol. 6, no. 4, pp. 201–208, 2000. ·
15. R. S. Mangoubi and D. F. X. Mathaisel, "Optimizing gate assignments at airport terminals," Transportation Science, vol. 19, no. 2, pp. 173–188, 1985. ·
16. A. Bolat, "Assigning arriving flights at an airport to the available

gates," Journal of the Operational Research Society, vol. 50, no. 1, pp. 23–34, 1999.

17. A. Bolat, "Models and a genetic algorithm for static aircraft-gate assignment problem," Journal of the Operational Research Society, vol. 52, no. 10, pp. 1107–1120, 2001.
18. A. Lim and F. Wang, "Robust airport gate assignment," in Proceedings of the 17th IEEE International Conference on Tools with Artificial Intelligence (ICTAI ‹05), pp. 74–81, November 2005. ·
19. S. Yan and C.-H. Tang, "A heuristic approach for airport gate assignments for stochastic flight delays," European Journal of Operational Research, vol. 180, no. 2, pp. 547–567, 2007.
20. "Stochastic programming," in Handbooks in Operations Research and Management Science, A. Ruszczynski and A. Shapiro, Eds., vol. 10, Elsevier, New York, NY, USA, 2003.
21. A. Ascó, Constructive and evolutionary algorithms for airport baggage sorting station and gate assignment problems [Ph.D. thesis], School of Computer Science, 2013.

CHAPTER 10

FINDING THE EFFICIENCY STATUS AND EFFICIENT PROJECTION IN MULTIOBJECTIVE LINEAR FRACTIONAL PROGRAMMING: A LINEAR PROGRAMMING TECHNIQUE

S. Morteza Mirdehghan and Hassan Rostamzadeh

Department of Mathematics, College of Sciences, Shiraz University, Shiraz, Iran

ABSTRACT

Multiobjective linear fractional programming (MOLFP) problems are the important problems with special structures in multiobjective optimization. In the MOLFP problems, the objective functions are linear fractional functions and the constraints are linear; that is, the feasible set is a polyhedron. In this paper, we suggest a method to identify the efficiency status of the feasible solutions of an MOLFP problem. By the proposed method, an efficient projection on the efficient space for an inefficient solution is obtained. The proposed problems are constructed in linear programming structures.

INTRODUCTION

Multiobjective programming (MOP) is a well-known research field in optimization and operations research. The multiobjective optimization problems have several objective functions and a set of feasible solutions. There are several special structures in multiobjective optimization problems. One class of them includes multiobjective linear fractional programming (MOLFP) problems which contain several linear fractional objective functions and a polyhedron as the feasible set.

There are many methods to find the efficient solutions of multiobjective optimization problems which are constructed based on iterative, scalarization, interactive, and other methods. If the objective functions or the constraints are not linear, then we should solve a mathematical programming problem which is not linear, but if the objective functions and constraints are linear, then the current approaches, usually, solve linear programming (LP) problems which are usually more desirable than nonlinear programming problems in the computationally efforts.

One of the well-known and the interesting methods to find the efficiency status of a feasible solution in MOP is Benson's method which was proposed by Benson [1]. This method receives an initial feasible solution of the MOP problem and identifies the efficiency status of that. If the feasible solution is not efficient, then this approach produces an efficient solution as a projection of the under-assessment inefficient solution. Another well-known method to find the efficient solution of an MOP problem is the weighted sum method. By this method the MOP problem is converted to an optimization problem with a single objective function. The weights of the objective functions of the MOP problems are specified by the decision-maker's point of view. If the weights are nonnegative, then the optimal solutions are weakly efficient and in particular if the weights are positive, then the optimal solutions are efficient. Besides the weighted sum approach, the -constraint method is also a well-known technique to solve MOP problems. There is no aggregation of objective functions; instead only one of the original objectives is optimized, while the others are transformed to constraints. This approach was introduced by Haimes et al. [2], and an extensive discussion can be found in Chankong and Haimes [3].

The objective functions of the MOLFP problems are fractional but the numerator and denominator of the objective functions are affine. To check the efficiency status of a feasible solution of an MOLFP problem by the traditional approaches, we should solve a fractional programming problem which is not desirable compared to solving a linear programming problem.

A linear fractional programming (LFP) problem includes a polyhedron as the feasible set and a fractional objective function whose numerator and denominator are affine functions. In other words, the numerator and denominator of the objective function are convex and concave simultaneously, but the whole of the objective function usually is not convex and concave, while it is quasi convex and quasi concave, simultaneously. So all local optimal solutions of an LFP problem are global optimal solutions. Moreover, if the LFP problem has an optimal solution, then it has an extreme optimal solution. To guarantee the existence of an optimal solution of an LFP problem, we suppose that the feasible set is bounded and the denominator of the objective function is positive or negative in the whole of the feasible set. Charnes and Cooper [4] have shown that an LFP problem can be solved by a linear programming technique; that is, the optimal solution of the constructed linear programming problem is the optimal solution of the LFP problem, where the sign of the denominator of the objective function is not changed in the whole of the feasible set. There are some researches to focus on solving LFP problems such as Gilmore and Gomory [5], Martos [6], Illés et al. [7], Tantawy [8], and Odior [9].

Because of the importance of solving MOLFP problems in multiobjective optimization, some articles have been published on the subject of solving MOLFP problems. Most of the techniques in the published researches are based on interactive, iterative, linearization, parametric, and decomposition methods. Kornbluth and Steuer [10] found a weakly efficient solution of an MOLFP problem based on a simplex based algorithm. Meteve and Gueorguieva [11] identified a weakly efficient solution using a nonlinear programming problem. Also Caballero and Hernández [12] proposed a controlled estimation method to find a set of weakly efficient solutions. Dinkelbach [13] solved an MOLFP problem using a parametric technique and this approach was extended by Almogy and Levin [14], Crouzeix et al. [15], Falk and

Palocsay [16], Tammer et al. [17], Skiscimi and Palocsay [18], and Schaible and Shi [19].

In the parametric technique of Tammer et al. [17], the parameters are calculated by solving some equations and the obtained feasible solution, necessarily, is not efficient solution. Most of parametric techniques extracted from Dinkelbach [13] cannot guarantee solving the MOLFP problems. Also some of approaches to solve MOLFP problems are based on iterative techniques such as Costa [20], Kuno [21], Phuong and Tuy [22], Dai et al. [23], Costa and Alves [24], Matejaš and Perić [25], and Valipour et al. [26]. Most of techniques for solving MOLFP problems try to find an efficient solution for an MOLFP problem.

In this paper, at first, we propose an approach which identifies the efficiency status of an arbitrary feasible solution of an MOLFP problem. Then, we propose an approach which not only identifies the efficiency status of an arbitrary feasible solution but also finds an efficient projection of an arbitrary feasible solution. In these two approaches, we construct linear programming problems regarding the MOLFP problem. Moreover, we show that in evaluating the under-assessment feasible solution by the proposed linear programming problem, in the first approach, if the optimal value is zero, then the under-assessment feasible solution is an efficient solution. In the second approach, the optimal solution of the linear programming problem is a weakly efficient solution. In particular, if the optimal solution of the proposed problem is unique, then the optimal solution is an efficient solution. The behavior and application of our proposed problems are similar to Benson's [1] problem and -constraint problem, proposed by Haimes et al. [2], for solving MOLFP problems. The differences among Benson's method, -constraint method, and our proposed methods are the structures of the suggested problems. Our proposed problems are linear programming problems, while Benson's and the ε-constraint problem, corresponding to an MOLFP problem, not only have nonlinear objective functions but also have nonlinear constraints.

In Section 2, we introduce some notions, definitions, and properties which are required in the main discussions. In Section 3, the main discussions and properties are presented. Two numerical examples to illustrate our approaches are presented in Section 4, and the final section is conclusion.

PRELIMINARIES

Linear Fractional Programming Problem

The linear fractional programming problems are modeled as follows:

$$\text{Min} \quad f(x) = \frac{cx+\alpha}{dx+\beta},$$

$$\text{s.t.} \quad x \in X = \{x \in R^n : Ax \le b,\ x \ge 0\}, \tag{14}$$

where X is a nonempty and bounded set, $b \in \mathrm{R}^m$ and c, $d \in \mathrm{R}^n$ are vectors of (known) coefficients, $A \in \mathrm{R}^{m\times n}$ is a (known) matrix of coefficients, and α, β are constant in R. Moreover, $dx + \beta$ ¹ 0 for all $x \in X$.

To solve LFP problem (1), usually, the method of Charnes and Cooper [4] is used. Let $t = 1/(dx + \beta)$ and $y = xt$; therefore, problem (1) can be converted to the following problem:

$$\begin{aligned} \text{Min} \quad & f(x) = cy + \alpha t, \\ \text{s.t.} \quad & dy + \beta t = 1, \\ & Ay - bt \le 0, \\ & y \ge 0,\ t \ge 0. \end{aligned} \tag{2}$$

Theorem 1. Let $(y*, t*)$ be an optimal solution of problem (2); then, $x* = y*/t*$ is an optimal solution of LFP problem (1).

Proof. See Charnes and Cooper [4].

Multiobjective Optimization

Consider the following Multiobjective programming problem:

$$\text{Min} \quad f(x) = \left(f_1(x), \ldots, f_p(x)\right)^t,$$

$$\text{s.t.} \quad x \in X, \tag{3}$$

where $X \subseteq R^n$ is the feasible set and $f_i(x)$, $i = 1, 2, \ldots, p$, are the objective functions. Usually, there exist conflicts among objective functions in their targets; then, usually, there does not exist any feasible solution of an MOP problem that optimizes all objective functions. Therefore, the notions of efficient solutions and weakly efficient solutions are introduced in MOP instead of optimal solutions.

Definition 2. A point $\bar{x} \in X$ is an efficient solution or paretooptimal solution of problem (3) if there is no other $\bar{x} \in X$ such that $(x) \leq fk(x)$ for all $k = 1, \ldots, p$ and $f_j(x) < f_j(\bar{x})$ for at least one $j \in \{1, \ldots, p\}$.

Definition 3. A point $\bar{x} \in X$ is a weakly efficient solution of problem (3) if there is no other $x \in X$ such that $f_k(x) < f_k(\bar{x})$ for all $k = 1, \ldots, p$.

Definition 4. A point $\bar{x} \in X$ is a strictly efficient solution of problem (3) if there is no other $x \in X$, ^{1} $\bar{x}$, such that $f_k(x) \leq f_k(\bar{x})$ for all $k = 1, \ldots, p$

Remark 5. Definitions 2, 3, and 4 imply that each strictly efficient solution is an efficient solution and each efficient solution is a weakly efficient solution but the reverses, necessarily, are not true.

There are many methods to find the efficient solutions in MOP. One of the most well-known methods to identify the efficiency status of a feasible solution was proposed by Benson [1] which is presented here in synopsis.

The idea and the aim of Benson's method are on identifying the efficiency status of a feasible solution of MOP. If the mentioned feasible solution is not efficient and then Benson's method finds an efficient solution of the MOP which dominates the mentioned inefficient feasible solution. This found efficient solution is called the projection of the inefficient solution on the efficient space. The problem proposed by Benson [1] maximizes the distance between the vector of objective functions corresponding to the under-assessment feasible solution (x^0) and the vectors of objective functions corresponding to all feasible solutions which dominate (x^0). The distance function in Benson's method is $\|\cdot\|_1$ (L_1 -norm).

By the above explanation, Benson's method is modeled as follows:

$$\begin{aligned} \text{Max} \quad & \sum_{k=1}^{p} l_k, \\ \text{s.t.} \quad & f_k(x) + l_k \le f_k\left(x^0\right), \quad k = 1, 2, \ldots, p, \\ & l_k \ge 0, \\ & x \in X. \end{aligned} \tag{4}$$

Theorem 6. If $x*$ is an optimal solution of problem (4), then $x*$ is an efficient solution of problem (3).

Proof. See Benson [1].

Theorem 7. If the optimal objective value of (4) equals zero, then the feasible solution $x^0 \in X$ is efficient.

Proof. See Benson [1].

Another technique to solve the multiobjective optimization problems is the ε-constraint method which was introduced by Haimes et al. [2]. In this method, only one of the original objectives is minimized, while the others are transformed to constraints. The ε-constraint problem associated with the MOP problem (3) is formulated as

$$\begin{aligned} \text{Min} \quad & f_j(x), \\ \text{s.t.} \quad & f_k(x) \le \varepsilon_k, \quad k = 1, 2, \ldots, p, \ k \ne j, \\ & x \in X, \end{aligned} \tag{5}$$

where $\varepsilon \in R^p$.

Theorem 8. Let $x*$ be an optimal solution of (5) for some j, and then $x*$ is a weakly efficient solution of (3).

Proof. See Haimes et al. [2].

Theorem 9. Let $x*$ be a unique optimal solution of (5) for some j, then $x*$ is a strictly efficient solution of (3), and therefore $x*$ is efficient.

Proof. See Haimes et al. [2].

THE PROPOSED APPROACHES

The multiobjective linear fractional programming problems are formulated as follows:

$$\text{Min} \quad f(x) = \left(\frac{c_1 x + \alpha_1}{d_1 x + \beta_1}, \ldots, \frac{c_p x + \alpha_p}{d_p x + \beta_p} \right)^t,$$

$$\text{s.t.} \quad x \in X = \{x \in R^n : Ax \leq b, \ x \geq 0\}, \tag{6}$$

where X is a nonempty and bounded set, $A \in R^{m \times n}$, $b \in R^m$, c_k, $d_k \in R^n$, and αk, $\beta_k \in R$ for all $k = 1, 2, \ldots, p$. Moreover, the feasible set X is a polytope (bounded polyhedron) and $d_k x + \beta_k$ is positive or negative in all feasible solutions of MOLFP problem (6) for all $k = 1, 2, \ldots, p$. Without loss of generality, in this paper we assume that $d_k x + \beta_k > 0$ for all $x \in X$ and $k = 1, 2, \ldots, p$.

To explain the proposed approaches, at first, we express Benson's method for MOLFP problem (6) and convert it to a linear programming problem. Benson's problem corresponding to MOLFP problem (6) is as follows:

$$\text{Max} \quad \sum_{k=1}^{p} l_k,$$

$$\text{s.t.} \quad \frac{c_k x + \alpha_k}{d_k x + \beta_k} + l_k \leq f_k\left(x^0\right), \quad k = 1, 2, \ldots, p,$$

$$l_k \geq 0, \quad k = 1, 2, \ldots, p,$$

$$x \in X, \tag{7}$$

where $x^0 \in X$ is a feasible solution of (6) and $f_k(x^0) = (c_k x^0 + \alpha_k)/(d_k x^0 + \beta_k)$, $k = 1, \ldots, p$

Let X ¹ 0 be a bounded set. It implies that the feasible set of (7) is bounded. Then, the efficient MOP problem (6) is not empty and problem (7) has an optimal solution.

Regarding the nonlinearity of the first group of the constraints of problem (7), we convert these constraints to linear constraints as follows:

$$\frac{c_k x + \alpha_k}{d_k x + \beta_k} + l_k \leq f_k\left(x^0\right) \Longrightarrow$$

$$\frac{c_k x + \alpha_k}{d_k x + \beta_k} - f_k\left(x^0\right) \leq -l_k \Longrightarrow$$

$$\frac{c_k x + \alpha_k - f_k\left(x^0\right)\left(d_k x + \beta_k\right)}{d_k x + \beta_k} \leq -l_k, \tag{8}$$

for $k = 1, 2, \ldots, p$. Since, for all $x \in X$, we have $d_k x + \beta_k > 0$ and $-l_k \leq 0$, $k = 1, 2, \ldots, p$, we have $c_k x + \alpha_k - f_k(x^0)(d_k x + \beta_k) \leq 0$ for $k = 1, 2, \ldots, p$.

Consider the following linear programming problem:

$$\text{Max} \quad \sum_{k=1}^{p} l'_k,$$

$$\text{s.t.} \quad c_k x + \alpha_k + l'_k \leq f_k\left(x^0\right)\left(d_k x + \beta_k\right),$$

$$k = 1, 2, \ldots, p, \tag{9}$$

The following theorem shows the relation between the optimal objective value of the above problem and the efficient solutions of MOLFP problem (6).

Theorem 10. If the optimal objective value of (9) equals zero, then the feasible solution x^0 is an efficient solution of MOLFP problem (6).

Proof. With respect to the constraints of problem (9) and because of $l'_k \geq 0$, $k = 1, 2, \ldots, p$, we conclude the following results for all x in the feasible set of (9):

$$c_k x + \alpha_k + l'_k \le f_k\left(x^0\right)\left(d_k x + \beta_k\right) \Longrightarrow$$

$$c_k x + \alpha_k \le f_k\left(x^0\right)\left(d_k x + \beta_k\right),$$

$$k = 1, 2, \ldots, p. \tag{10}$$

Because of $d_k x+\beta_k > 0$ for all $x \in X$, we have $(c_k x+\alpha_k)/(d_k x+ \beta_k) \le f_k(x^0)$ for all $k = 1, 2, \ldots, p$.

Since all coefficients of variables l'_k, $k = 1, 2, \ldots, p$, in the objective function of problem (9) are positive, they are equal to one, and $x*$ is an optimal solution of (9); therefore, l'_k associated with $x*$, which is shown by l'^*_k, satisfies in the following relations:

$$c_k x^* + \alpha_k + l'^*_k = f_k\left(x^0\right)\left(d_k x^* + \beta_k\right) \Longrightarrow$$

$$-l'^*_k = c_k x^* + \alpha_k - f_k\left(x^0\right)\left(d_k x^* + \beta_k\right),$$

$$k = 1, 2, \ldots, p, \tag{11}$$

and so

$$-\sum_{k=1}^{p} l'^*_k = \sum_{k=1}^{p}\left(c_k x^* + \alpha_k - f_k\left(x^0\right)\left(d_k x^* + \beta_k\right)\right). \tag{12}$$

Now, suppose that $(x*, l'^*) = (x^*, 0)$ is an optimal solution of (9) and by contradiction suppose that x^0 is not an efficient solution of MOLFP problem (6). Then, there exists a feasible solution $\bar{x} \in X$ such that $(c_k \bar{x} + \alpha_k)/(d_k \bar{x} + \beta_k) \le (c_k x^0 + \alpha_k)/(d_k x^0 + \beta_k) = f_k(x^0)$ for all k and $(c_j \bar{x} + \alpha_j)/(d_j \bar{x} + \beta_j) < (c_j x^0 + \alpha_j)/(d_j x^0 + \beta_j) = f_j(x^0)$ for some j.

Let $I = \{k \mid (c_k \bar{x} + \alpha_k)/(d_k \bar{x} + \beta_k) = f_k(x^0)\}$ and $l'_k = f_k(x^0)(d_k \bar{x} + \beta_k) - (c_k \bar{x} + \alpha_k)$ for all $k = 1, 2, \ldots, p$. These relations imply that $l'_k = 0$ for all $k \in I$ and $l'_k > 0$ for all $k \notin I$. Because of $\bar{l}'_k \ge 0$, $k = 1, 2, \ldots, p$, we conclude that $(\bar{x}, l')$ is a feasible solution of (9).

Also, $\{1, 2, \ldots, p\} \setminus I^{1} 0$ and we have $\sum_{k=1}^{p} \bar{l}'_k > 0$ which contradicts with the assumption of the optimality of $(x*, l'*) = (x*, 0)$ in (9). Because the objective value of (9) corresponding to $(\bar{x}, \bar{l}')$ is strictly positive while the objective value of (9) corresponding to (x^*, l'^*) is zero, therefore $x*$ is an efficient solution of (6).

Now, we express the ε-constraint method for MOLFP problem (6). Let x^0 be a feasible solution of (6) and $\varepsilon_k = (x^0)$, $k = 1, 2, \ldots, p$. Therefore, ε-constraint problem (5) corresponding to MOLFP problem (6) is as follows:

$$
\begin{aligned}
\text{Min} \quad & \frac{c_j x + \alpha_j}{d_j x + \beta_j}, \\
\text{s.t.} \quad & Ax \leq b, \\
& \frac{c_k x + \alpha_k}{d_k x + \beta_k} \leq f_k\left(x^0\right), \quad k = 1, \ldots, p, \ k \neq j, \\
& x \geq 0.
\end{aligned} \tag{13}
$$

We transform the above nonlinear programming problem to a linear programming problem using Charnes and Cooper's transformation method. Let $t = 1/(d_j x + \beta_j)$ and $y = xt$. The above problem is transformed to the following linear programming problem:

$$
\begin{aligned}
\text{Min} \quad & c_j y + \alpha_j t, \\
\text{s.t.} \quad & Ay - bt \leq 0, \\
& c_k y + \alpha_k t \leq f_k\left(x^0\right)\left(d_k y + \beta_k t\right), \\
& \qquad\qquad k = 1, \ldots, p, \ k \neq j, \\
& d_j y + \beta_j t = 1, \\
& y \geq 0, \ t \geq 0.
\end{aligned} \tag{14}
$$

Theorem 11. Let $(y*, t*)$ be an optimal solution of problem (14) for some j, and then $x* = y*/t*$ is an optimal solution of problem (13).

Proof. At first, we claim that t is not zero in all feasible solutions of (14). By contradiction, suppose that$(y, t) = (\bar{y}, 0)$ is a feasible solution of (14). So $A\bar{y} \leq 0$ and $\bar{y} \geq 0$.These imply that $\bar{y}$ is a recession direction of X (the feasible set of MOLFP (6)) which is a contradiction to the primary assumption of the boundedness of X. So $t>0$ in all feasible solutions, and in particular t is positive in all optimal solutions of (14).

$(y*, t*)$ is an optimal solution of (14) and $t* > 0$. Let $x* = y*/t*$. We prove that $x*$ is an optimal solution of (13).

$$Ax^*t^* = Ay^* \leq bt^* \Longrightarrow$$

$$Ax^* \leq b. \tag{15}$$

Moreover

$$c_k x^* t^* + \alpha_k t^* = c_k y^* + \alpha_k t^* \leq f_k\left(x^0\right)\left(d_k y^* + \beta_k t^*\right)$$

$$= f_k\left(x^0\right)\left(d_k x^* t^* + \beta_k t^*\right), \tag{16}$$

for all $k = 1, 2, \ldots, p$ and k ¹ j. These imply that

$$\frac{c_k x^* + \alpha_k}{d_k x^* + \beta_k} \leq f_k\left(x^0\right), \quad k = 1, \ldots, p, \ k \neq j. \tag{17}$$

So $x*$ is a feasible solution of problem (13). We show that $x*$ is an optimal solution of (13). By contradiction, suppose that $x*$ is not an optimal solution of (13). Therefore, there exists feasible solution $\bar{x}$ of (13) such that

$$\frac{c_j \bar{x} + \alpha_j}{d_j \bar{x} + \beta_j} < \frac{c_j x^* + \alpha_j}{d_j x^* + \beta_j}. \tag{18}$$

Let $\bar{t} = 1/(d_j\bar{x} + \beta_j)$ and $\bar{y} = \bar{x}\bar{t}$. It is easy to see that $(\bar{y}, \bar{t})$ is a feasible solution of (14). Moreover, from (18) we have $c_j y + \alpha_j \bar{x} < c_j y* + \alpha_j \bar{t}*$, which contradicts with the assumption that $(y*, t*)$ is an optimal solution of (14). Therefore $x* = y*/t*$ is an optimal solution of problem (13), and the proof is completed.

Theorem 12. Let $(y*, t*)$ be an optimal solution of problem (14) for some j, and then $x* = y*/t*$ is a weakly efficient solution for MOLFP problem (6). In particular, if $(y*, t*)$ is a unique optimal solution of (14), then $x* = y*/t*$ is a strictly efficient solution and an efficient solution for MOLFP problem (6).

Proof. Suppose that $(y*, t*)$ is an optimal solution of (14). Therefore, by Theorem 11, $x* = y*/t*$ is an optimal solution of (13) and, regarding to Theorem 8, $x*$ is weakly efficient solution of MOLFP problem (6). Moreover, if $(y*, t*)$ is a unique optimal solution of problem (14), then $x* = y*/t*$ is a unique optimal solution of problem (13) and, regarding to Theorem 9, $x*$ is a strictly efficient solution and, therefore, it is an efficient solution of MOLFP problem (6), which completes the proof.

Remark 13. By Theorem 12, we conclude that an optimal solution of problem (14) clarifies a weakly efficient solution of MOLFP problem (6). If the optimal solution of (14) is unique, then this optimal solution determines an efficient solution of MOLFP problem (14). Otherwise, if problem (14) has alternative optimal solutions, then to clarify the efficiency status of an optimal solution of (14) we use problem (9) and Theorem 10.

EXAMPLES

Example 1. Consider the following MOLFP problem:

$$\begin{aligned}
\text{Min} \quad & f_1(x) = \frac{x_1 + 2x_2 + 5}{x_1 + x_2 + 2}, \\
\text{Min} \quad & f_2(x) = \frac{x_1 + x_2 + 10}{-x_1 + 2x_2 + 4}, \\
\text{s.t.} \quad & x_1 \leq 3, \\
& x_2 \leq 4, \\
& x_1, x_2 \geq 0.
\end{aligned} \tag{19}$$

According to problem (14) for j=1, problem (19) can be converted to the following problem:

$$\begin{aligned} \text{Min} \quad & y_1 + 2y_2 + 5t, \\ \text{s.t.} \quad & y_1 + y_2 + 10t \le f_2\left(x^0\right)\left(-y_1 + 2y_2 + 4t\right), \\ & y_1 + y_2 + 2t = 1, \\ & y_1 - 3t \le 0, \\ & y_2 - 4t \le 0, \\ & y_1, y_2, t \ge 0, \end{aligned} \tag{20}$$

where x^0 is a feasible solution of problem (19).

For each arbitrary feasible point, which is called an underassessment feasible solution, we will solve problem (20) and obtain the projection of the under-assessment feasible solution of problem (19) on the weakly efficient space of that. Then, we check the efficient status of the projection solution using problem (9) and Theorem 10. The optimal solutions of problem (20) corresponding to several under-assessment feasible solutions and their efficiency statuses have been listed in Table 1. Note that if $(y*, t*)$ is a unique optimal solution of problem (20), then $y*/t*$ is an efficient solution for MOLFP problem (19) and we do not need to check the projection by (9). Figure 1 shows the feasible set and the efficient solutions of (19). The rectangle with vertex A, B, C, O and its interior is the set of feasible solutions. All points on bold segments AB and BC are the efficient solutions of (19).

Table 1: Results of Example 1 using different feasible solutions of (19)

Row	Feasible solution (x^0)	Efficiency status of x^0	Projection
1	A = (0.0000, 4.0000)	Efficient	(0.0000, 4.0000)
2	B = (3.0000, 4.0000)	Efficient	(3.0000, 4.0000)
3	C = (3.0000, 0.0000)	Efficient	(3.0000, 0.0000)
4	(2.1078, 4.0000)	Efficient	(2.1078, 4.0000)
5	(0.0000, 0.0000)	Inefficient	(3.0000, 2.6250)
6	(0.0000, 2.0000)	Inefficient	(1.6000, 4.0000)
7	(1.6789, 1.5678)	Inefficient	(3.0000, 2.7424)

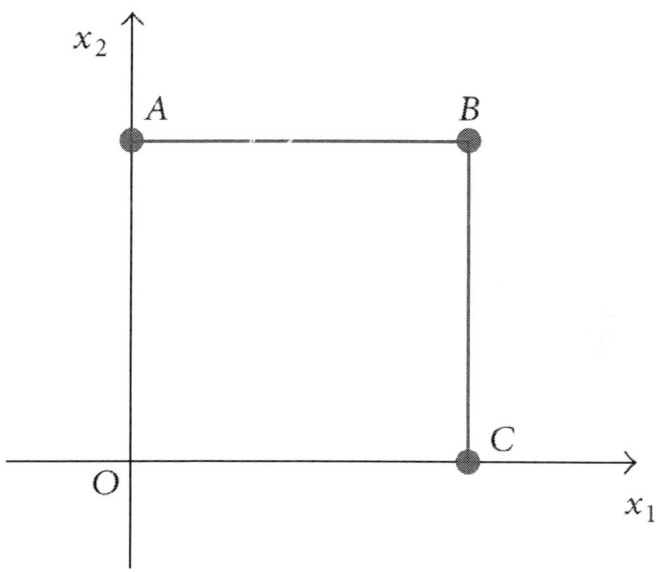

A = (0, 4)

B = (3, 4)

C = (3, 0)

Figure 1: Feasible and efficient sets of problem (19).

Note that points A, B, and C are extremely efficient. If we choose them as under-assessment points (x^0) and solve problem (20), then the optimal solutions of (20) corresponding to the three points, which are the projections of these feasible solutions on the efficient space, are unique and equal to themselves.

Table 1 presents the results of examining 7 feasible solutions of (19) of which 6 feasible solutions are on the boundary of the feasible set and one of them is an interior point of the feasible set. The projection points corresponding to the feasible points in rows 1 to 4 are themselves and they are efficient by problem (9) which are also endorsed by Figure 1. In particular, the points in rows 1 to 3 are basic feasible solutions of (19) and so they are the efficient extreme points of (19). Also the feasible solution in row 4 is on segments AB.

The other feasible solutions in rows 5 to 7 are inefficient. The points in rows 5 and 6 are on the inefficient boundary of the feasible set of (19) and their projections are on segments BC and AB, respectively. The point in row 7 is an interior point of the feasible set of (19) and its projection is on segment BC.

Example 2. Here, we consider an MOLFP problem with three objective functions which has been extracted from Kornbluth and Steuer [10] as follows:

$$\begin{aligned}
\text{Min} \quad & f_1(x) = \frac{-x_1 + 4}{-x_2 + 3}, \\
\text{Min} \quad & f_2(x) = \frac{x_1 - 4}{x_2 + 1}, \\
\text{Min} \quad & f_3(x) = x_1 - x_2, \\
\text{s.t.} \quad & -x_1 + 4x_2 \leq 0, \\
& x_1 - 0.5x_2 \leq 4, \\
& x_1, x_2 \geq 0.
\end{aligned} \tag{21}$$

The feasible set of the above MOLFP problem has been shown in Figure 2. The feasible set is a triangle with vertices A, B, and C. All the bold segments are efficient except points B and D which are weakly efficient. The other points in the feasible set are inefficient. For more illustration, see Kornbluth and Steuer [10].

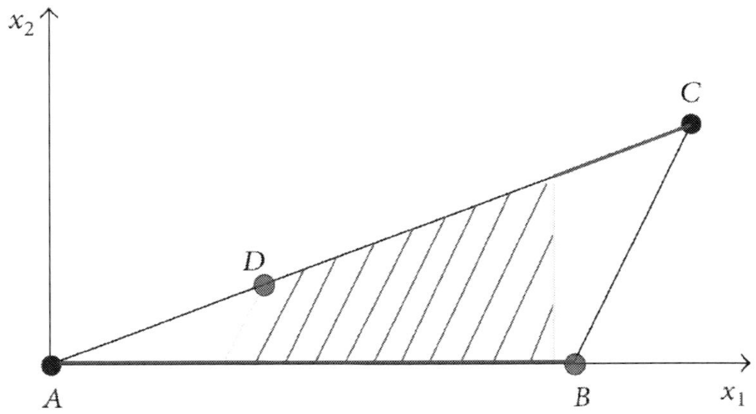

$A = (0, 0)$ $\quad$ $C = (4.5714, 1.1429)$

$B = (4, 0)$ $\quad$ $D = (1.3, 0.3)$

Figure 2: Feasible and efficient regions of problem (21).

Regarding (6), (13), and (14) for *j*=2, MOLFP problem (21) is transformed to the following linear programming problem:

$$\begin{aligned}
\text{Min} \quad & y_1 - 4t, \\
\text{s.t.} \quad & -y_1 + 4t \le f_1\left(x^0\right)\left(-y_2 + 3t\right), \\
& y_1 - y_2 \le f_3\left(x^0\right)t, \\
& y_2 + t = 1, \\
& -y_1 + 4y_2 \le 0, \\
& y_1 - 0.5y_2 - 4t \le 0, \\
& y_1, y_2, t \ge 0,
\end{aligned} \tag{22}$$

where x^0 is a feasible solution of problem (21).

By solving linear programming problem (22) for each arbitrary feasible solution of (21), we can obtain a projection of x^0 on the weakly efficient space of (21). The results of solving problem (21) for several feasible solutions have been listed in Table 2. In Table 2, we see that columns 2 and 4 corresponding to rows 1 to 4 are the same. In other words, the projections of the under-assessment feasible solutions in rows 1 to 4 are themselves and they are efficient using problem (9) corresponding to (21). Furthermore, all other feasible solutions in rows 5 to 8 are inefficient and their projections, obtained from (22), are efficient.

Table 2: The projections of several feasible solutions of (21) and their efficiency statuses

Row	Feasible point (x^0)	Efficiency status of x^0	Projection
1	$A = (0.0000, 0.0000)$	Efficient	(0.0000, 0.0000)
2	$C = (4.5714, 1.1429)$	Efficient	(4.5714, 1.1429)
3	(4.0000, 1.0000)	Efficient	(4.0000, 1.0000)
4	(1.5558, 0.0000)	Efficient	(1.5558, 0.0000)
5	$B = (4.0000, 0.0000)$	Inefficient (Weakly efficient)	(4.0000, 0.0706)
6	$D = (1.3000, 0.3000)$	Inefficient (Weakly efficient)	(1.0000, 0.0000)
7	(0.5000, 0.0625)	Inefficient	(0.4255, 0.0000)
8	(0.7690, 0.1168)	Inefficient	(0.6381, 0.0000)

CONCLUSION

In real world, many of events and problems are modeled as multiobjective programming problems. Multiobjective linear fractional programming (MOLFP) problems have special structures among multiobjective programming problems. The feasible sets of them are polyhedron and also their objective functions are fractional whose nominators and denominators are affine functions.

Because of the special structure of the MOLFP problems, in this paper we suggest a linear programming technique to find the efficiency status of a feasible solution of an MOLFP problem, and if it is not efficient, then we project it on the efficient space of the MOLFP problem. For this purpose, we propose two linear programming problems. One of the proposed linear programming problems attempts to find a feasible solution of the MOLFP problem which dominates the under-assessment feasible solution. If there does not exist any optimal solution of the proposed linear programming problem which equals the under-assessment feasible solution, then the under-assessment feasible solution is inefficient. In this case the optimal solution is a projection of the under-assessment feasible solution on the weakly efficient set of the MOLFP problems. Otherwise, if the optimal solution of the proposed problem is the same as the under-assessment feasible solution, then the feasible solution is weakly efficient. Another proposed linear programming problem identifies the efficiency statuses of the feasible solutions. We can check the efficiency status of an arbitrary feasible solution or a projection point, obtained by another problem, by this linear programming problem.

REFERENCES

1. H. P. Benson, "Existence of efficient solutions for vector maximization problems," Journal of Optimization Theory and Applications, vol. 26, no. 4, pp. 569–580, 1978.
2. Y. Y. Haimes, L. S. Lasdon, and D. A. Wismer, "On a bicriterion formulation of the problems of integrated system identification and system optimization," IEEE Transactions on Systems, Man, and Cybernetics, vol. 1, pp. 296–297, 1971.
3. V. Chankong and Y. Haimes, Multiobjective Decision Making: Theory and Methodology, Elsevier Science, New York, NY, USA, 1983.
4. A. Charnes and W. W. Cooper, "Programming with linear fractional functionals," Naval Research Logistics, vol. 9, pp. 181–186, 1962.
5. P. C. Gilmore and R. E. Gomory, "A linear programming approach to the cutting stock problem. Part II," Operations Research, vol. 11, no. 6, pp. 863–888, 1963.
6. B. Martos, "Hyperbolic programming," Naval Research Logistics, vol. 11, pp. 135–155, 1964.
7. T. Illés, Á. Szirmai, and T. Terlaky, "The finite criss-cross method for hyperbolic programming," European Journal of Operational Research, vol. 114, no. 1, pp. 198–214, 1999.
8. S. F. Tantawy, "A new procedure for solving linear fractional programming problems," Mathematical and Computer Modelling, vol. 48, no. 5-6, pp. 969–973, 2008.

9. A. O. Odior, "An approach for solving linear fractional programming problems," International Journal of Engineering & Technology, vol. 1, pp. 298–304, 2012.
10. J. S. H. Kornbluth and R. E. Steuer, "Multiple objective linear fractional programming," Management Sciences, vol. 27, no. 9, pp. 1024–1039, 1981.
11. B. Metev and D. Gueorguieva, "A simple method for obtaining weakly efficient points in multiobjective linear fractional programming problems," European Journal of Operational Research, vol. 126, no. 2, pp. 386–390, 2000.
12. R. Caballero and M. Hernández, "The controlled estimation method in the multiobjective linear fractional problem," Computers & Operations Research, vol. 31, no. 11, pp. 1821–1832, 2004.
13. W. Dinkelbach, "On nonlinear fractional programming," Management Science, vol. 13, pp. 492–498, 1967.
14. Y. Almogy and O. Levin, "A class of fractional programming problems," Operations Research, vol. 19, pp. 57–67, 1971.
15. J.-P. Crouzeix, J. A. Ferland, and S. Schaible, "An algorithm for generalized fractional programs," Journal of Optimization Theory and Applications, vol. 47, no. 1, pp. 35–49, 1985.
16. J. E. Falk and S. W. Palocsay, "Image space analysis of generalized fractional programs," Journal of Global Optimization, vol. 4, no. 1, pp. 63–88, 1994.
17. K. Tammer, C. Tammer, and E. Ohlenderf, "Multicriterial fractional optimization," in Parametric Optimization and Related Topics IV, J. Guddat, H. T. Jongen, F. Nozicka, G. Still, and F. Twilt, Eds., pp. 359–370, Peter Lang, Berlin, Germany, 1997.
18. C. C. Skiscim and S. W. Palocsay, "Minimum spanning trees with sums of ratios," Journal of Global Optimization, vol. 19, no. 2, pp. 103–120, 2001.
19. S. Schaible and J. Shi, "Recent developments in fractional programming: single ratio and maxmin case," in Proceedings of the 3rd International Conference in Nonlinear Analysis, W. Takahashi and T. Tanaka, Eds., pp. 493–506, Yokohama Publisher, Yokohama, Japan, 2004.

20. J. P. Costa, "Computing non-dominated solutions in MOLFP," European Journal of Operational Research, vol. 181, no. 3, pp. 1464–1475, 2007.
21. T. Kuno, "A branch-and-bound algorithm for maximizing the sum of several linear ratios," Journal of Global Optimization, vol. 22, no. 1–4, pp. 155–174, 2002.
22. N. T. H. Phuong and H. Tuy, "A unified monotonic approach to generalized linear fractional programming," Journal of Global Optimization, vol. 26, no. 3, pp. 229–259, 2003.
23. Y. Dai, J. Shi, and S. Wang, "Conical partition algorithm for maximizing the sum of DC ratios," Journal of Global Optimization, vol. 31, no. 2, pp. 253–270, 2005.
24. J. P. Costa and J. M. Alves, "A reference point technique to compute nondominated solutions in MOLFP," Journal of Mathematical Sciences, vol. 161, no. 6, pp. 820–831, 2009.
25. J. Matejaš and T. Perić, "A new iterative method for solving multiobjective linear programming problem," Applied Mathematics and Computation, vol. 243, pp. 746–754, 2014.
26. E. Valipour, M. A. Yaghoobi, and M. Mashinchi, "An iterative approach to solve multiobjective linear fractional programming problems," Applied Mathematical Modelling, vol. 38, no. 1, pp. 38–49, 2014.

CHAPTER 11

GENERALIZED MEAN FOR ROBUST PRINCIPAL COMPONENT ANALYSIS

JiyongOh, NojunKwak

Graduate School of Convergence Science and Technology, AICT, Seoul National University, 1 Gwanak-ro, Gwanak-gu, Seoul 08826, Republic of Korea

ABSTRACT

In this paper, we propose a robust principal component analysis (PCA) to overcome the problem that PCA is prone to outliers included in the training set. Different from the other alternatives which commonly replace L_2-norm by other distance measures, the proposed method alleviates the negative effect of outliers using the characteristic of the generalized mean keeping the use of the Euclidean distance. The optimization problem based on the generalized mean is solved by a novel method. We also present a generalized sample mean, which is a generalization of the sample mean, to estimate a robust mean in the presence of outliers. The proposed method shows better or equivalent performance than the conventional PCAs in various problems such as face reconstruction, clustering, and object categorization.

Keywords: Generalized mean, Principal component analysis, Robust PCA, Dimensionality reduction

INTRODUCTION

Dimensionality reduction [1] is a classical problem in pattern recognition and machine learning societies, and numerous methods have been proposed to reduce the data dimensionality. Principal component analysis (PCA) [2] is one of the most popular unsupervised dimensionality reduction methods which tries to find a subspace where the average reconstruction error of the training data is minimized. It is useful in representation of input data in a low dimensional space and it has been successfully applied to face recognition [3,4], visual tracking [5], clustering [6,7], and so on.

When automatically collecting a large data set, outliers may be contained in the collected data since it is very difficult to examine whether each sample of data is outlier or not [8]. It is well known that, in this case, the conventional PCA is sensitive to outliers because it minimizes the reconstruction errors of training data in terms of the mean squared error and a few outliers with large errors dominate the objective function. This problem has been addressed in many studies [8–16]. Among them, some studies utilized L_1-norm instead of L_2-norm in the formulation of optimization problem to improve the robustness of PCA against outliers [9–11]. In [9], the cost function for optimization was constructed based on L_1-norm and a convex programming was employed to solve the problem. R_1-PCA [10] was presented to obtain a solution with the rotational invariance, which is a fundamental desirable property for learning algorithms [17]. In [11], PCA-L_1 was proposed, which maximizes an L_1 dispersion in the reduced space and an extension of PCA-L_1 using L_p-norm with arbitrary p was also proposed in [14]. Other method utilizing L_p-norm was also presented in [15]. On the other hand, some of robust PCAs were recently developed using information theoretic measures [12,13]. He et al. [12] proposed MaxEnt-PCA which finds a subspace where Renyi's quadratic entropy [18] is maximized. Renyi's entropy was estimated by a non-parametric Parzen window technique. In [13], HQ-

PCA was developed based on the maximum correntropy criterion [19].

In this paper, we propose a new robust PCA method based on the power mean or the generalized mean [20], which can become the arithmetic, geometric, and harmonic means depending on the value of its parameter. The proposed method, *PCA-GM*, is a generalization of the conventional PCA by replacing the arithmetic mean with the generalized mean. The proposed method can effectively prevent outliers from dominating objective function by controlling the parameter in the generalized mean. Moreover, it is rotational invariant because it still uses the Euclidean distance as the distance measure between data samples. In doing so, we also propose a generalized sample mean, which is an enhancement of the conventional algebraic sample mean against outliers to address the problem that the sample mean is easily affected by outliers. It is used in the proposed PCA-GM instead of the arithmetic mean. The optimization problems based on the generalized mean are efficiently solved using a mathematical property of the generalized mean. Recently, Candés et al. proposed a robust PCA [21], which is sometimes referred to as RPCA in the literature, where data matrix is tried to be represented as a sum of a low rank matrix, which corresponds to reconstructions of data, and a sparse matrix, which corresponds to reconstruction errors different from the methods mentioned above. It can model pixel-wise noise effectively using the sparse matrix, thus it has been known that RPCA is useful in the applications such as background modeling from surveillance video and removing shadows and specularities from face images [21] by using each element in the reconstruction error vector (the column of the sparse matrix). On the other hand, in this paper, we will utilize distance metric in removing the effect of outliers like the previously mentioned methods, and an entire sample is considered as an outlier if it has a large norm of the reconstruction error vector.

The remainder of this paper is organized as follows. Section 2 briefly introduces PCA and the state-of-the-art robust PCAs. The proposed method is described in Section 3. It is demonstrated in Section 4 that the proposed method gives better performances in face reconstruction and clustering problems than other variants of PCA. Finally, Section 5 concludes this paper.

PCA AND ROBUST PCAS

Let us consider a training set of N n-dimensional samples {lbracesupsubxi} $^{N}_{i}$=1. Assuming that the samples have zero-mean, PCA is to find an orthonormal projection matrix W∈$R^{n\times m}$||(m||n) by which the projected samples {yi=W^Txi}$^{N}_{i}$=1 have the maximum variance in the reduce space. It is formulated as follows:

$$\mathbf{W}_{PCA} = \arg\max_{\mathbf{W}} \operatorname{tr}(\mathbf{W}^T\mathbf{S}\mathbf{W}),$$

where $\mathbf{S} = \frac{1}{N}\sum_{i=1}^{N}\mathbf{x}_i\mathbf{x}_i^T$ is a sample covariance matrix and tr(A) is the trace of a square matrix A. The projection matrix WPCA can be also found from the viewpoint of projection errors, i.e., it minimizes the average of the squared projection errors or reconstruction errors. Mathematically, it is represented as the optimization problem minimizing the following cost function:

$$J_{L_2}(\mathbf{W}) = \frac{1}{N}\sum_{i=1}^{N} \|\mathbf{x}_i - \mathbf{W}\mathbf{W}^T\mathbf{x}_i\|_2^2,$$

where $\|x\|_2$ is the L_2-norm of a vector x. The two optimization problems are equivalent and easily solved by obtaining the m eigenvectors associated with the m largest eigenvalues of S. Although PCA is simple and powerful, it is prone to outliers [8,9] because J_{L2}(W) is based on the mean squared reconstruction error. To learn a subspace robust to outliers, Ke and Kanade [9] proposed to minimize an L_1-norm based objective function as follows:

$$J_{L_1}(\mathbf{W}) = \frac{1}{N}\sum_{i=1}^{N} \|\mathbf{x}_i - \mathbf{W}\mathbf{W}^T\mathbf{x}_i\|_1,$$

where $\|x\|_1$ is the L_1-norm of a vector x. They also present an iterative method to obtain the solution for minimizing J_{L1}(W).

Although L_1-PCA minimizing J_{L1}(W) can relieve the negative effect of outliers, it is not invariant to rotations. In [10], Ding et al. proposed R_1-PCA, which is rotational invariant, at the same time is robust to outliers. It is to minimize the following objective function:

$$J_{R_1}(\mathbf{W}) = \sum_{i=1}^{N} \rho\left(\sqrt{\mathbf{x}_i^T\mathbf{x}_i - \mathbf{x}_i^T\mathbf{W}\mathbf{W}^T\mathbf{x}_i}\right),$$

where ρ(·) is a generic loss function and the Cauchy function or Huber's M-estimator [22] was used for ρ(·) in [10]. Huber's M-estimator ρH(s) is defined as

$$\rho_H(s) = \begin{cases} s^2 & \text{if } |s| \leq c, \\ 2c|s| - c^2 & \text{otherwise} \end{cases} \tag{1}$$

where c is the cutoff parameter that controls the regularization effect of weights in a weighted covariance matrix. Note that ρH(s) becomes a quadratic or a linear function of |s| depending on the value of s. The solution for minimizing JR1(W) was obtained by performing a subspace iteration algorithm [23].

On the other hand, PCA-L_1 was developed in [11] motivated by the duality between maximizing variance and minimizing reconstruction error. It maximizes an L_1dispersion among the projected samples, $\sum_{i=1}^{N} \|\mathbf{W}^T\mathbf{x}_i\|_1$. A novel and efficient method for maximizing the L_1 dispersion was also presented in [11]. The method allows PCA-L_1 to be performed by much less computational effort than R_1-PCA.

HQ-PCA is formulated based on the maximum correntropy criterion in terms of information theoretic learning. Without the zero-mean assumption, which is necessary in other variants of PCA, HQ-PCA maximizes the correntropy estimated between a set of training samples $\{\mathbf{x}_i\}_{i=1}^{N}$ and the set of their reconstructed samples $\{\mathbf{W}\mathbf{y}_i + \mathbf{m}\}_{i=1}^{N}$, where m is a data mean. Mathematically, HQ-PCA tries to maximize the following objective function:

$$\arg\max_{\mathbf{W},\mathbf{m}} \sum_{i=1}^{N} g\left(\sqrt{\bar{\mathbf{x}}_i^T\bar{\mathbf{x}}_i - \bar{\mathbf{x}}_i^T\mathbf{W}\mathbf{W}^T\bar{\mathbf{x}}_i}\right), \tag{2}$$

where g(x)=exp(−x²/2σ²) is the Gaussian kernel and $\bar{\mathbf{x}}_i = \mathbf{x}_i - \mathbf{m}$. Note that HQ-PCA finds a data mean as well as a projection matrix. Using the Welsch M-estimator $\rho_W(x)=1-g(x)$, HQ-PCA is regarded as a robust M-estimator formulation because it is equivalent to finding WH and mH that minimize the following objective function:

$$J_{HQ}(\mathbf{W},\mathbf{m}) = \sum_{i=1}^{N} \rho_W\left(\sqrt{\bar{\mathbf{x}}_i^T\bar{\mathbf{x}}_i - \bar{\mathbf{x}}_i^T\mathbf{W}\mathbf{W}^T\bar{\mathbf{x}}_i}\right). \tag{3}$$

In [13], the optimization problem in (2) was effectively solved in the half-quadratic optimization framework, which is often used to address nonlinear optimization problems in information theoretic learning.

ROBUST PRINCIPAL COMPONENT ANALYSIS BASED ON GENERALIZED MEAN

Generalized Mean for Positive Numbers

For a p≠0, the generalized mean or power mean Mp of {a_i>0,i=1,…,N} [20]is defined as

$$\mathcal{M}_p\{a_1,\ldots,a_N\} = \left(\frac{1}{N}\sum_{i=1}^{N} a_i^p\right)^{1/p}.$$

The arithmetic mean, the geometric mean, and the harmonic mean are special cases of the generalized mean when p=1,p→0, and p=−1, respectively. Furthermore, the maximum and the minimum values of the numbers can also be obtained from the generalized mean by making p→∞ and p→−∞, respectively. Note that as p decreases (increases), the generalized mean is more affected by the smaller (larger) numbers than the larger (smaller) ones, i.e., controlling p makes it possible to adjust the contribution of each number to the generalized mean. This characteristic is useful in the situation where data samples should be differently handled according to their importance, for example, when outliers are contained in the training set.

In [24], it was shown that the generalized mean of a set of positive numbers can be expressed by a nonnegative linear combination of the elements in the set and, in this paper, it is further simplified as follows:

$$\sum_{i=1}^{K} a_i^p = b_1 a_1 + \cdots + b_K a_K$$
$$b_i = a_i^{p-1}, \quad i = 1, \ldots, K. \tag{4}$$

Note that each weight b_i has the same value of 1 if p=1, where the generalized mean becomes the arithmetic mean. It is also noted that, if p

is less than one, the weight b_i increases as a_i decreases. This means that, when p<1, the generalized mean is more influenced by the small numbers in $\{a_i\}_{i=1}^{K}$, and the extent of the influence increases as p decreases. This equation plays an important role in solving the optimization problems using the generalized mean.

Generalized Sample Mean

Most conventional PCAs commonly assume that training samples have zero-mean. To satisfy this assumption, all of the samples are subtracted by the sample mean, i.e., $x_i - m_S$ for i=1,…,N, where $\mathbf{m}_S = \frac{1}{N}\sum_{i=1}^{N}\mathbf{x}_i$. The conventional sample mean can be considered as the center of the samples in the sense of the least square, i.e.,

$$\mathbf{m}_S = \arg\min_{\mathbf{m}} \frac{1}{N}\sum_{i=1}^{N} \|\mathbf{x}_i - \mathbf{m}\|_2^2. \quad (5)$$

In (5), a small number of outliers in the training samples dominate the objective function because the objective function in (5) is constructed based on the squared distances. To obtain a robust sample mean in the presence of outliers, a new optimization problem is formulated by replacing the arithmetic mean in (5) with the generalized mean as

$$\mathbf{m}_G = \arg\min_{\mathbf{m}} \left(\frac{1}{N}\sum_{i=1}^{N} \left(\|\mathbf{x}_i - \mathbf{m}\|_2^2\right)^p\right)^{1/p}.$$

This problem is equivalent to (5) if p=1. As mentioned in the previous subsection, the contribution of a large number to the objective function decreases as p decreases. Thus, the negative effect of outliers can be alleviated if p<1. From now on, we will call mG as the *generalized sample mean*. Using the fact that x^p with p>0 being a monotonic increasing function of x for x>0, this problem can be converted to

$$\mathbf{m}_G = \arg\min_{\mathbf{m}} \sum_{i=1}^{N} \left(\|\mathbf{x}_i - \mathbf{m}\|_2^2\right)^p. \quad (6)$$

Although the minimization in (6) should be changed into the maximization when p<0, we only consider positive values of p in this paper.

The necessary condition for mG to be a local minimum is that the gradient of the objective function in (6) with respect to m is equal to zero, i.e.,

$$\frac{\partial}{\partial \mathbf{m}} \sum_{i=1}^{N} \left(\| \mathbf{x}_i - \mathbf{m} \|_2^2 \right)^p = 0.$$

However, it is hard to find a closed-form solution of the above equation. Although any gradient-based iterative algorithms can be applied to obtain mG, they usually have slow convergence speed. Alternatively, we develop a novel method based on (4), which is more efficient than gradient-based iterative methods. Our method for solving the problem in (6) is an iterative one, similar to the expectation–maximization algorithm [25].

In the derivation, we decompose (6) into the form of (4) and consider the weight b_i in (4) as a constant. Then, (6) can be approximated by a quadratic function of $\|x_i - m\|_2$ which can easily be optimized. The details are as follows. Let us denote the value of m after t iterations as $m^{(t)}$. The first step of the update rule is, for m close to a fixed m(t), to represent the objective function in (6) as a linear combination of $\|x_i - m^{(t)}\|_{22}$ using (4), i.e.,

$$\sum_{i=1}^{N} \left(\| \mathbf{x}_i - \mathbf{m} \|_2^2 \right)^p \approx \sum_{i=1}^{N} \alpha_i^{(t)} \| \mathbf{x}_i - \mathbf{m} \|_2^2,$$

where

$$\alpha_i^{(t)} = \left(\| \mathbf{x}_i - \mathbf{m}^{(t)} \|_2^2 \right)^{p-1}. \tag{7}$$

Here, the approximation becomes exact when $m = m^{(t)}$. Note that the objective function near $m^{(t)}$ can be approximated as a quadratic function of m without computing the Hessian matrix of the objective function. The next step is to find $m^{(t+1)}$ that minimizes the approximated function based on the computed $\alpha_i^{(t)}$, i.e.,

$$\frac{\partial}{\partial \mathbf{m}} \sum_{i=1}^{N} \alpha_i^{(t)} \| \mathbf{x}_i - \mathbf{m} \|_2^2 = 0.$$

The solution of this equation is just the weighted average of the samples as follows:

$$\mathbf{m}^{(t+1)} = \frac{1}{\sum_{j=1}^{N} \alpha_j^{(t)}} \sum_{i=1}^{N} \alpha_i^{(t)} \mathbf{x}_i. \tag{8}$$

This update rule with the two steps is repeated until a convergence condition is satisfied. This procedure is summarized in Algorithm 1.

1: Input: $\{x_1,\ldots,x_N\}$, p>0.
2: t←0.
3: $m^{(t)} \leftarrow m_S$.
4: repeat
5: Approximation: For fixed m(t), **compute** α1(t),…,αN(t) **according to (7)**.
6: Minimization: Using the computed α1(t),…,αN(t), **update** $m^{(t+1)}$ **according to (8)**.
7: t←t+1.
8: until A stop criterion is satisfied
9: Output: $m_G = m^{(t)}$.

Algorithm 1: Generalized sample mean.

Note that a weighted average is computed at each iteration in Algorithm 1. Thus, it can be said that Algorithm 1 is a special case of the mean shift algorithm [26]. It is also noted that the number of initial points is only one, which is set to m_S. Since non-convex optimization methods depend on initial points, they are generally conducted several times started from different initial points and the solution is selected as the one providing the best performance. However, we have empirically found that Algorithm 1 started from m_S converges to a local optimum point that is enough robust to outliers.

To demonstrate the robustness of the generalized sample mean obtained by Algorithm 1, we randomly generated 100 samples from a two-dimensional Gaussian distribution with the mean mi=0 and covariance matrix Σ_i=diag[0.5,0.5] for inliers and also generated 10 samples from another two-dimensional Gaussian distribution with the mean $m_o=[5,5]^T$ and covariance matrix Σ_o=diag[0.3,0.3] for outliers. Using the generated samples, the sample mean was computed and two generalized sample means were also obtained by Algorithm 1 with p=0.1 and p=0.2, respectively. Figure 1 shows the arithmetic sample mean and the two generalized sample means together with the generated samples. It is obvious that the generalized sample means are located close to the mean of the inliers, [0,0]T, whereas the arithmetic sample mean is much more biased by the ten outliers. This illustrates that the generalized sample mean with an appropriate value of p is more robust to outliers than the arithmetic sample mean.

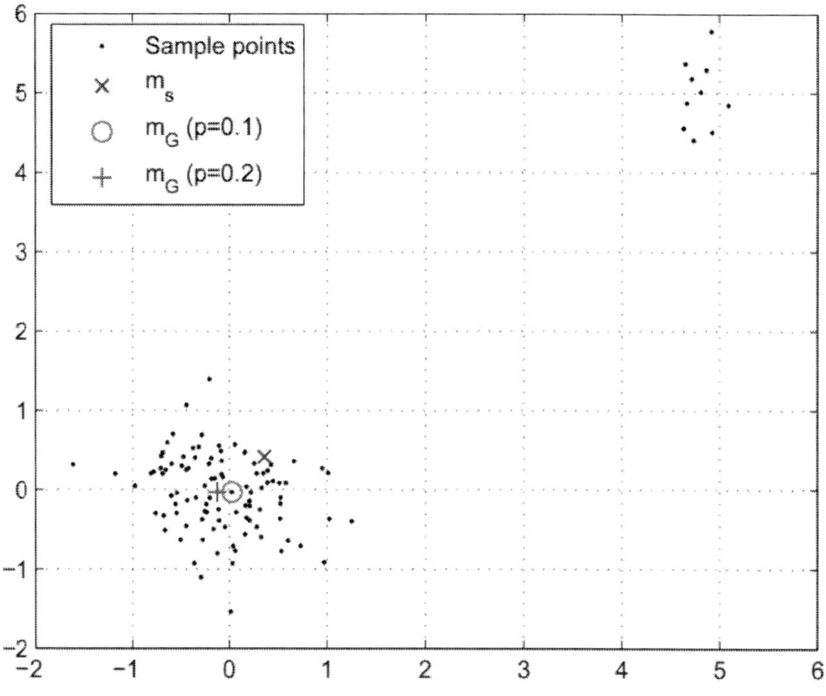

Figure 1. 2D toy example with 100 inliers and 10 outliers. The arithmetic mean (m_S) and the generalized sample mean (m_G) are marked.

Principal Component Analysis Using Generalized Mean

For a projected sample W^Tx, the squared reconstruction error e(W) can be computed as

$$e(\mathbf{W}) = \tilde{\mathbf{x}}^T\tilde{\mathbf{x}} - \tilde{\mathbf{x}}^T\mathbf{W}\mathbf{W}^T\tilde{\mathbf{x}},$$

where $\tilde{x}=x-m$. We use the generalized sample mean m_G for m. To prevent outliers corresponding to large e(W) from dominating the objective function, we propose to minimize the following objective function:

$$J_G(\mathbf{W}) = \left(\frac{1}{N}\sum_{i=1}^{N}[e_i(\mathbf{W})]^p\right)^{1/p}, \tag{9}$$

where $e_i(\mathbf{W}) = \tilde{\mathbf{x}}_i^T\tilde{\mathbf{x}}_i - \tilde{\mathbf{x}}_i^T\mathbf{W}\mathbf{W}^T\tilde{\mathbf{x}}_i$ is the squared reconstruction error of xi with respect to W. Note that $J_G(W)$ is formulated by replacing the arithmetic mean in $J_{L2}(W)$ with the generalized mean keeping the use of the Euclidean distance and it is equivalent to $J_{L2}(W)$ if p=1. The negative effect raised by outliers is suppressed in the same way as in (6). Also, the solution that minimizes $J_G(W)$ is rotationally invariant because each $e_i(W)$ is measured based on the Euclidean distance. To obtain W_G, we develop an iterative optimization method similar to Algorithm 1.

Like the optimization problem for m_G in the previous subsection, under the assumption that p>0, the optimization problem based on (9) is firstly converted as follows:

$$\begin{aligned}\mathbf{W}_G &= \underset{\mathbf{W}^T\mathbf{W}=\mathbf{I}}{\arg\min}\left(\frac{1}{N}\sum_{i=1}^{N}[e_i(\mathbf{W})]^p\right)^{1/p} \\ &= \underset{\mathbf{W}^T\mathbf{W}=\mathbf{I}}{\arg\min}\sum_{i=1}^{N}[e_i(\mathbf{W})]^p,\end{aligned} \tag{10}$$

Next, let us denote $W^{(t)}$ as the value of $W \in R^{n \times m}$ after the *t*-th iteration. Near a fixed $W^{(t)}$, the converted objective function in (10) can be approximated as a quadratic function of W according to (4) as

$$\sum_{k=1}^{N}[e_i(\mathbf{W})]^p \approx \sum_{i=1}^{N}\beta_i^{(t)}e_i(\mathbf{W}),$$

where

$$\beta_i^{(t)} = \left[e_i(\mathbf{W}^{(t)})\right]^{p-1}. \tag{11}$$

Here, the approximation becomes exact if $W=W^{(t)}$. After calculating each $\beta_i^{(t)}$, $W^{(t+1)}$ can be computed by minimizing the approximated function as

$$\begin{aligned}\mathbf{W}^{(t+1)} &= \arg\min_{\mathbf{W}} \sum_{i=1}^{N} \beta_i^{(t)} e_i(\mathbf{W}) \\ &= \arg\max_{\mathbf{W}} \mathrm{tr}\left(\mathbf{W}^T \mathbf{S}_\beta^{(t)} \mathbf{W}\right),\end{aligned} \tag{12}$$

where $\mathbf{S}_\beta^{(t)} = \sum_{i=1}^{N} \beta_i^{(t)} \tilde{\mathbf{x}}_i \tilde{\mathbf{x}}_i^T$.

Input: $\{x_1,\ldots,x_N\}$, m_G, ***m*, *p*.**
$t \leftarrow 0$.
$W^{(t)} \leftarrow W_{PCA} \in R^{n\times m}$.
repeat
Approximation: For fixed $W^{(t)}$, compute $\beta_1^{(t)},\ldots,\beta_N^{(t)}$ using (11).
Minimization: Using the computed $\beta_1^{(t)},\ldots,\beta_N^{(t)}$, find $W^{(t+1)}$ by solving the eigenvalue problem in (12).
$t \leftarrow t+1$.
until A stop criterion is satisfied
Output: $W_G = W^{(t)}$.

Algorithm 2: PCA-GM.

Note that $S_\beta^{(t)}$ is a weighted covariance matrix and the columns of $W^{(t+1)}$ are the *m* orthonormal eigenvectors associated with the largest *m* eigenvalues of $S_\beta^{(t)}$. These two steps are repeated until a convergence criterion is satisfied. Algorithm 2summarizes this iterative procedure of computing W_G. Unfortunately, the update rule in Algorithm 2 does not guarantee that $J_G(W^{(t+1)})<J_G(W^{(t)})$. Nonetheless, the experimental results show that W_G obtained by the algorithm is good enough.

To help understanding of Algorithm 2, we made another toy example as shown in Figure 2(a) where 110 two dimensional samples are plotted. Among the samples, 100 samples are regarded as inliers and the others are regarded as outliers. The samples were generated as the following rule:

$x_i \sim N(0,1)$,

$y_i = x_i + \epsilon_i$,

where the random noise ϵ_i is sampled from N(0,0.5^2) for inliers and N(0,3^2) for outliers, respectively. Figure 2(b) shows the objective function of PCA-GM in (9) with p=0.3 for the samples as shown in Figure 2(a). We can see from Figure 2(b) that the conventional PCA is prone to the ten outliers because its objective function is minimized around W=[cos60°sin60°]T. However, PCA-GM is robust to the outliers because its objective function is minimized at W=[cos48.9°sin48.9°]T, which is close to the solution without the outliers W*=[cos45°sin45°]T. Given an initial projection vector W(0)=[cos30°sin30°]T, the approximation step in Algorithm 2 gives a quadratic function corresponding to the red dashed line in Figure 2(b). In the second step, the next iteration $W^{(1)}$ is determined as [cos32.1°sin32.1°]T by minimizing the approximate function. Interestingly, it can be said that the approximate function plays a similar role of an upper bound of the objective function around $W^{(0)}$ in this update rule. It is also noted that the approximated function at the local optimal point W=[cos48.9°sin48.9°]T has its minimum as the same location, which is denoted as the magenta dashed dotted line in Figure 2(b). This means that Algorithm 2 converges to the local minimum point of the objective function for the problem shown in Figure 2(a).

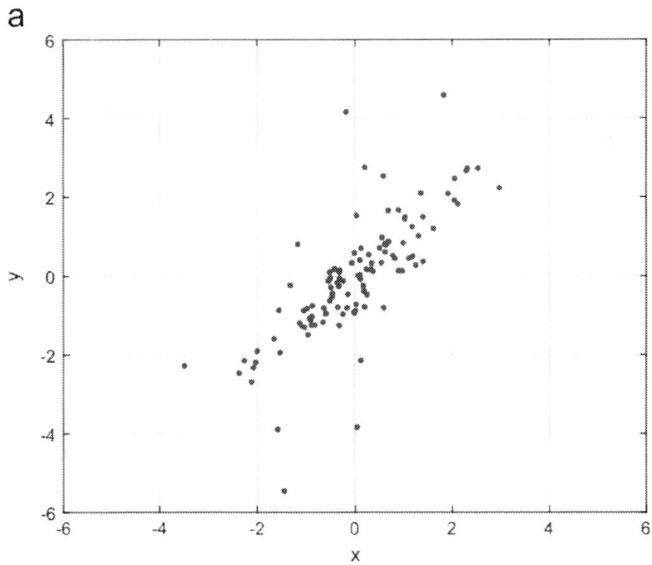

b

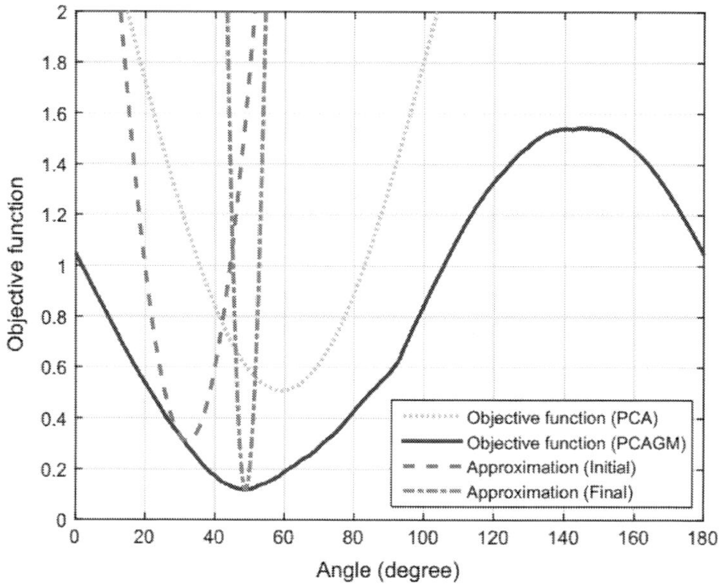

Figure 2. (a) A toy example to illustrate Algorithm 2. (b) The values of the objective functions $J_{L2}(W)$, $J_G(W)$ are plotted. The quadratic approximations of $J_G(W)$ at the initial point $W^{(0)}$=[cos30°sin30°]T and the final point $W^{(t)}$=[cos48.9°sin48.9°]T are also plotted.

In practice, when $e_i(W^{(t)})$ is zero or very small for any *i*, $[e_i(W^{(t)})]p-1$ is numerically unstable if p<1, and Algorithm 2 cannot proceed anymore. This problem can also occur in Algorithm 1. It can be overcome by adding a small constant δ into each $e_i(W)$ as

$$e_i(\mathbf{W})' = \tilde{\mathbf{x}}_i^T\tilde{\mathbf{x}}_i - \tilde{\mathbf{x}}_i^T\mathbf{W}\mathbf{W}^T\tilde{\mathbf{x}}_i + \delta, \quad (13)$$

where δ should be small enough that the modified objective function is not affected too much. This perturbation also changes $\mathbf{S}_\beta^{(t)}$ in (12) into $\widehat{\mathbf{S}}_\beta^{(t)}$ as

$$\widehat{\mathbf{S}}_\beta^{(t)} = \sum_{i=1}^{N} \beta_i^{(t)} \left(\tilde{\mathbf{x}}_i \tilde{\mathbf{x}}_i^T + \frac{\delta}{n} \right),$$

where *n* is the original dimensionality of data.

EXPERIMENTS

To evaluate the proposed method, we considered face reconstruction, digit clustering, and object categorization problems, the first two of which were addressed in [11,13], respectively. The proposed method was compared with PCA, PCA-L_1, R_1-PCA, and HQ-PCA. Except for the conventional PCA, they have the parameters to be predetermined and we determined the values of the parameters according to the recommendations in [10,11,13]. Also, in PCA-GM, the generalized sample mean was used instead of the sample mean, and the perturbation parameter δ in (13) was set to 0.01 times the minimum of e_i(WPCA) for i=1,…,N. For the iterative algorithms as R_1-PCA, HQ-PCA, PCA-GM, the number of iterations was limited to 100.

Face Reconstruction

We collected 800 facial images from the subset 'fa' of the Color FERET database [27] for the face reconstruction problem. Each face image was normalized to a size of 40×50 pixels using the eye coordinates, which were obtained in the database. We simulated two types of outliers. For the first type of outliers, some of the facial images were randomly selected, and each of the selected images was occluded by a rectangular area, each pixel in which was randomly set to 0 (black) or 255 (white). The size and location of the rectangular area were randomly determined. Figure 3 shows examples of original normalized faces in the upper row and their corresponding faces occluded by the rectangular noise in the lower row. To evaluate the proposed method with different noise levels, we selected 80, 160, and 240 images from the 800 facial images and occluded them by rectangular noise, so that we made three training sets including 80, 160, and 240 occluded images. For the second type of outliers, other three training sets were constructed by adding 80, 160, and 240 dummy images (outlier) with the same size to the original 800 face images (inlier), so that the numbers of inliers and outliers in the three training sets are (800,80), (800,160), and (800,240) respectively. Each pixel in the dummy images was also randomly set to 0 or 255.

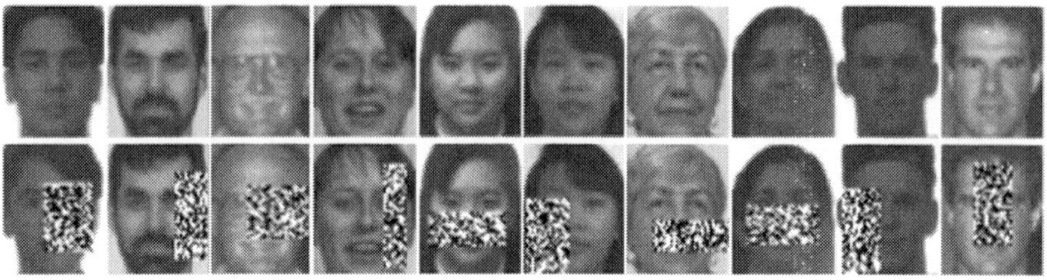

Figure 3. Examples of original face images (upper row) and the corresponding images (lower row) occluded by rectangular noise.

After applying different versions of PCAs to the training sets with the various numbers of extracted features m from 5 to 100, we compared the average reconstruction errors as in [11] defined as

$$\frac{1}{N}\sum_{i=1}^{N}\left\|\left(\mathbf{x}_i^{ori}-\mathbf{m}\right)-\mathbf{W}\mathbf{W}^T(\mathbf{x}_i-\mathbf{m})\right\|_2, \tag{14}$$

where x_i^{ori} and x_i are the i-th original unoccluded image and the corresponding training image, respectively, N is the number of the face images, and m is the mean of the original normalized faces. For the training sets related to the second type of outliers, the dummy images were excluded when measuring the average reconstruction errors, and x_i^{ori} and x_i were identical. Note that W is the projection matrix obtained from PCA, PCA-L_1, R_1-PCA, HQ-PCA, and PCA-GM for the various values of m. Moreover, PCA-GM was performed using 0.1, 0.2, 0.3, and 0.4 for the value of p to figure out the effect of it.

Figs. 4 and 5 show the average reconstruction errors measured as in (14) for the training sets constructed to simulate two types of outliers when 5≤m≤100. As shown in the figures, PCA-GM and HQ-PCA generally gave better performances than PCA, PCA-L_1, and R_1-PCA regardless of the types of outliers and the level of noise, and they yielded competitive results to each other. When the number of the occluded images is 240, which corresponds to Figure 4(c), HQ-PCA provided lower average reconstruction errors than PCA-GM for m≤ 40 while PCA-GM with p=0.1 and p=0.2 gave better performances than HQ-PCA for m≥ 60. When the number of the dummy images is 80, which corresponds to Figure 5(a), the lower reconstruction errors could be obtained by PCA-GM rather than HQ-PCA when m≤ 60 while HQ-PCA performed better than PCA-GM for 80 ≤m< 100.

a

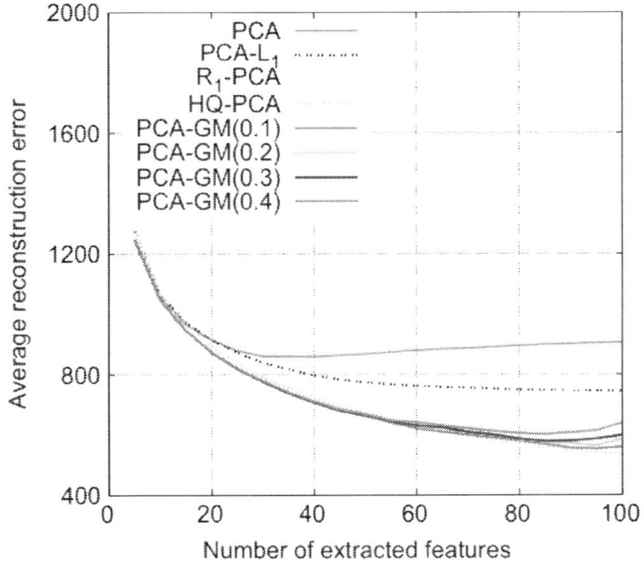
PCA
PCA-L_1
R_1-PCA
HQ-PCA
PCA-GM(0.1)
PCA-GM(0.2)
PCA-GM(0.3)
PCA-GM(0.4)
Average reconstruction error
2000
1600
1200
800
400
0
20
40
60
80
100
Number of extracted features

b

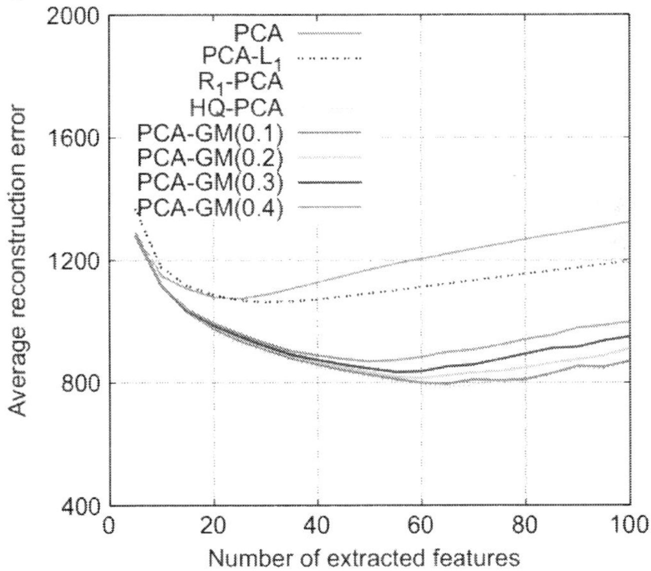
PCA
PCA-L_1
R_1-PCA
HQ-PCA
PCA-GM(0.1)
PCA-GM(0.2)
PCA-GM(0.3)
PCA-GM(0.4)
Average reconstruction error
2000
1600
1200
800
400
0
20
40
60
80
100
Number of extracted features

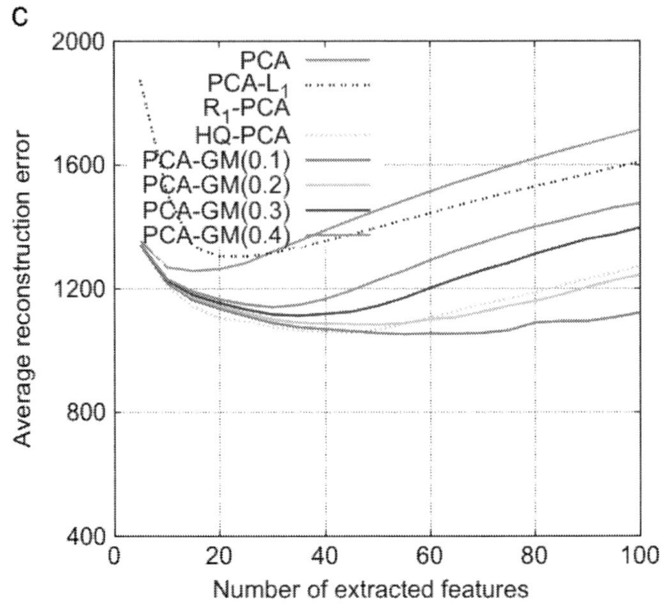

Figure 4. Average reconstruction errors of different PCA methods for the data sets where the numbers of inliers and outliers (occlusion) are (a) (720, 80), (b) (640,160), and (c) (560, 240).

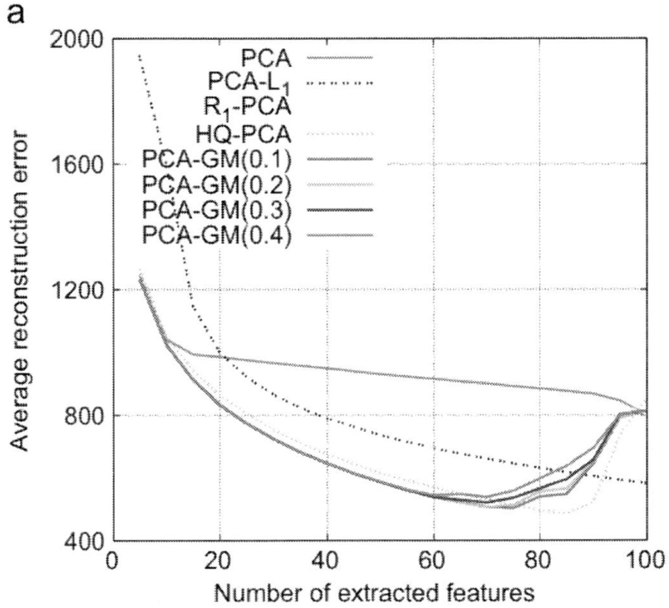

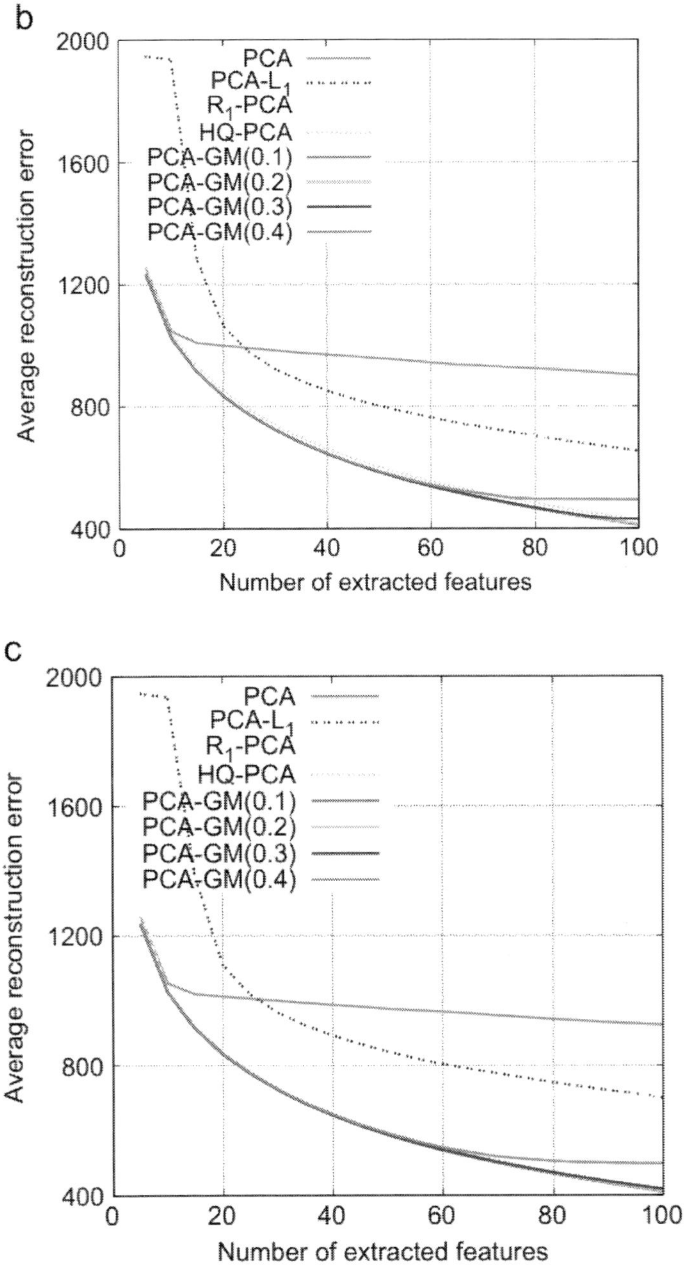

Figure 5. Average reconstruction errors of different PCA methods for the data sets where the numbers of inliers and outliers (dummy images) are (a) (800, 80), (b) (800, 160), and (c) (800, 240).

The effectiveness of the proposed method can also be found by visualizing projection matrices in terms of the *Eigenfaces*[3]. Figure 6 shows the first ten of Eigenfaces obtained by different PCA methods when m=40 and the number of outliers is 240 for both types of outliers. We can see that the Eigenfaces of HQ-PCA and PCA-GM are less contaminated from the outliers than PCA, PCA-L_1, and R_1-PCA. Also, it can be seen from the figure that PCA-L_1 yielded projection matrix with a different property. This may be a reason of the fact that PCA-L_1 provided relatively large reconstruction errors when m is small as shown in Figs. 4 and 5(c).

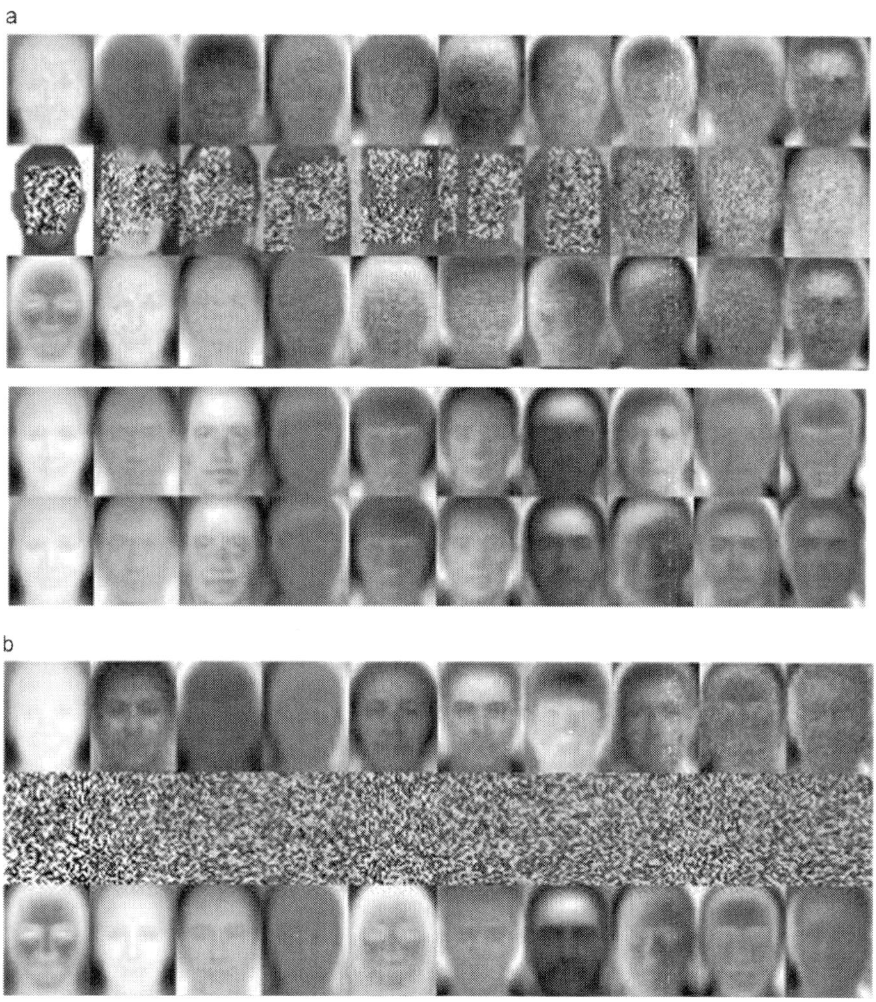

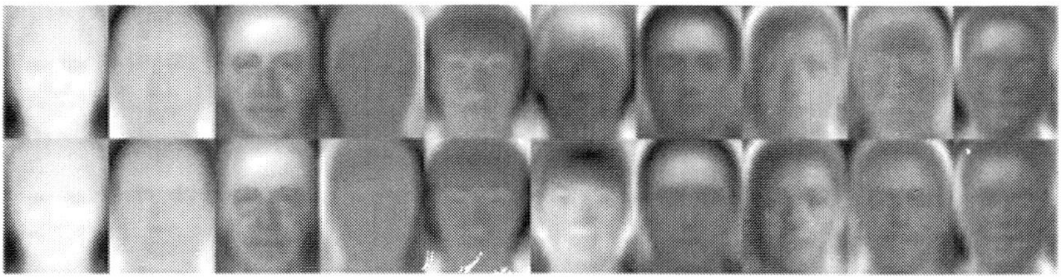

Figure 6. Eigenfaces obtained by PCA, PCA-L_1, R_1-PCA, HQ-PCA, and PCA-GM with p=0.1 in order of row. (a) Occlusion noise. (b) Dummy image noise.

The effect of p in PCA-GM was as expected. For the occlusion noise as shown in Figure 4, the lower values of p gave better performances and the performance differences are more distinct as m and noise level increase. For the dummy images as shown in Figure 5, PCA-GM showed almost similar performances for all the values of p except for 0.4 when m≥ 70. These results agree with the fact that the generalized mean of a set of positive numbers depends on small numbers more and more as p gets smaller.

Clustering

The clustering problem was dealt with using a subset of the MNIST handwritten digit database, which includes a training set of 60,000 examples and a test set of 10,000 examples. We randomly gathered 100 examples per the digits '3', '8', and '9' from the first 10,000 examples in the training set. To simulate outliers, we also randomly gathered 60 examples corresponding to the other digits from the same 10,000 examples. Thus, our training set for the clustering problem consists of 300 inliers and 60 outliers, which were normalized to unit norm. Figure 7 shows nine images of the inliers in the upper row and nine images of the outliers in the lower row.

Figure 7. Examples of MNIST handwritten digit images used as inliers (first row; 3, 8, 9) and outliers (second row; other digits).

After obtaining projection matrices by applying various versions of PCAs to the training set, *K*-means clustering with K=3 was performed using the projected inlier examples. For the initial means of the *K*-means clustering, we selected the two examples with the largest distance and then selected another example which had the largest sum of the distances from the previously selected two examples. The clustering accuracy was computed based on the class labels assigned to the examples in the database. Table 1 shows the clustering accuracy for various numbers of extracted features. As the previous experiments, we conducted PCA-GM using the settings of $p \in \{0.1, 0.2, 0.3, 0.4\}$. The best performance was achieved when p=0.3 which is reported in Table 1. Considering the clustering accuracy without the dimensionality reduction was 70%, PCA-GM improved the clustering accuracy by 5%. Different from the results of the face reconstruction problem in the previous subsection, R_1-PCA and PCA-L_1 gave similar highest clustering accuracy as PCA-GM, while HQ-PCA performed poorly than PCA-GM. However, R_1-PCA and PCA-L_1provided the highest accuracy when m=300 whereas PCA-GM yielded the best performance when m=200.

Table 1. Clustering accuracy (%) of the digit images corresponding to '3', '8', and '9' in the reduced spaces which are obtained from the training set containing the other digit images as outliers

m	PCA	PCA-L_1	R_1-PCA	HQ-PCA	PCA-GM
50	70.00	69.00	69.67	70.00	70.00
100	70.00	72.00	70.00	69.33	69.67
150	70.67	74.00	70.67	70.00	70.00
200	70.33	73.67	70.33	73.67	**75.00**
250	70.33	73.67	73.67	74.00	74.00
300	70.33	**75.00**	**75.00**	73.67	73.67

Object Categorization

We evaluated the proposed method by performing object categorization on Small NORB data set [28]. The NORB data set consists of images of 50 different objects belonging to 5 categories each of which contains 10 objects. For each category, the images of 5 objects shown in Figure 8(a) belong to its training set and those of 5 objects shown in Figure 8(b) belong to its test set. The Small NORB data set is a subset of the NORB data set comprising 24,300 images for training and 24,300 images for testing, which are normalized with the size of 96 × 96 pixels on uniform background. Each object in the data set was captured under 18 azimuths, 9 elevations, and 6 light conditions. To evaluate the proposed method for different numbers of training samples, we uniformly sampled the three image capture variables to construct three training sets with 3375, 12,150, and 24,300 samples. We also resized the images in the data set to 48 × 48 and 64 × 64 pixels for computational efficiency. Consequently, we have six training sets with different number of samples (N) and dimensionality of input samples (n).

a

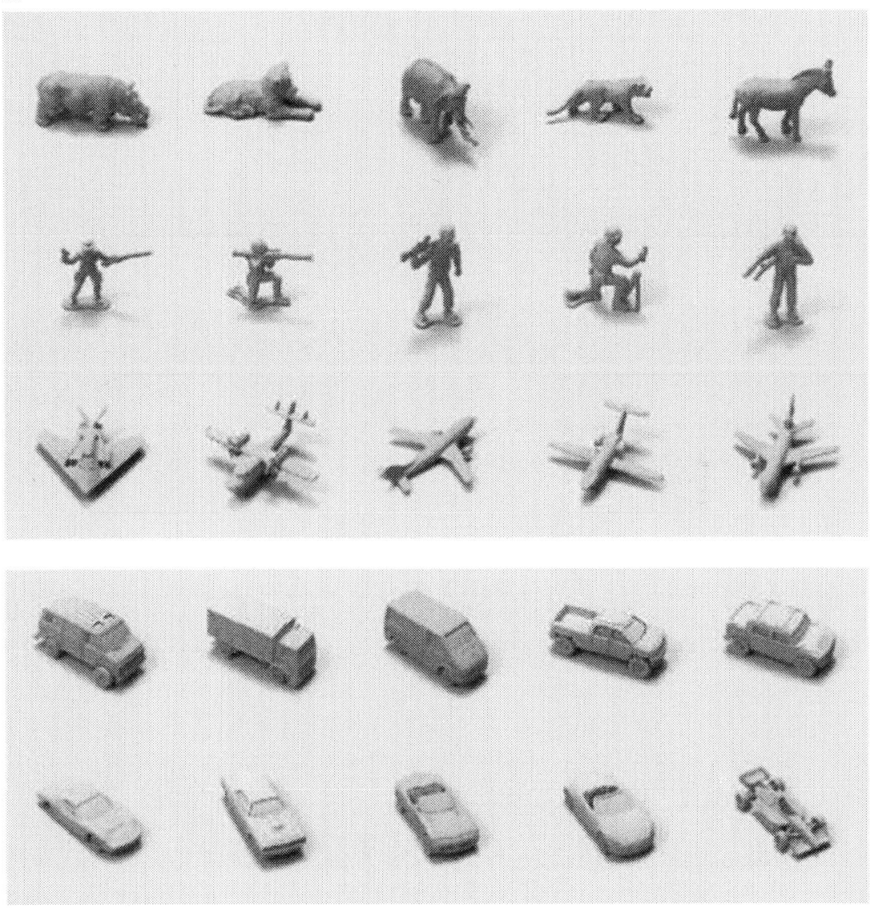

b

Figure 8. Images of objects in (a) training and (b) test sets of NORB data set.

Although there are various approaches to categorize an arbitrary sample z corresponding to an image of an object, we performed the categorization by the *nearest-to-subspace*, i.e., z is determined to belong to the category minimizing the distance from z to the subspace spanned by the training samples in the category. For the distance from z to the subspace of the *i*-th category, we employed the squared residual error of z to the subspace computed as $\tilde{\mathbf{z}}_i^T \tilde{\mathbf{z}}_i - \tilde{\mathbf{z}}_i^T \mathbf{W}_i \mathbf{W}_i^T \tilde{\mathbf{z}}_i$, where $\tilde{\mathbf{z}}_i = \mathbf{z} - \mathbf{m}_i$ and Wi is the orthonormal basis of the subspace, which corresponds to the projection matrix and can be obtained by one of the PCA methods aforementioned. Also, m_i is the mean of the training samples in the *i*-th category. We used the sample mean m_S for mi in PCA, PCA-L_1, and R_1-

PCA while we used m_H and m_G instead of m_S in HQ-PCA and PCA-GM, respectively. For the purpose of comparison, the categorization accuracy was evaluated varying the dimensionality of subspaces (m) from 5 to 50.

Figure 9 shows the categorization accuracy measured on the 24,300 test images in Small NORB data set. It is necessary to note that artificial outliers were not used in this experiments different from the previous ones. We can see that PCA-GM with an appropriate value of p is competitive with the conventional PCA when N=3375 and the proposed method provides higher categorization accuracies than PCA as N increases. Especially when N=24,300, the proposed method achieves the best performance for all the cases of m. This trend appears in both cases of n=48×48and n=64×64. However, the other variants of PCA did not gave higher accuracies than the conventional PCA for most cases. In particular, HQ-PCA, which showed competitive performance in the face reconstruction experiments, resulted in the lowest categorization accuracy. This means that the proposed method can be an effective alternative to PCA in object categorization using the nearest-to-subspace when training data is enough.

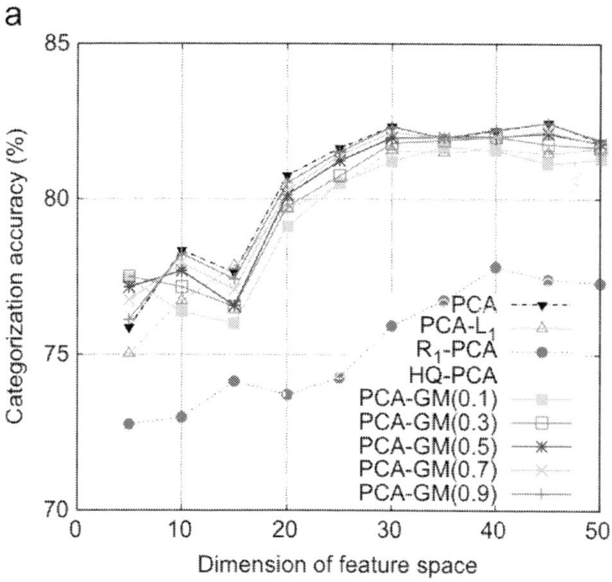

b

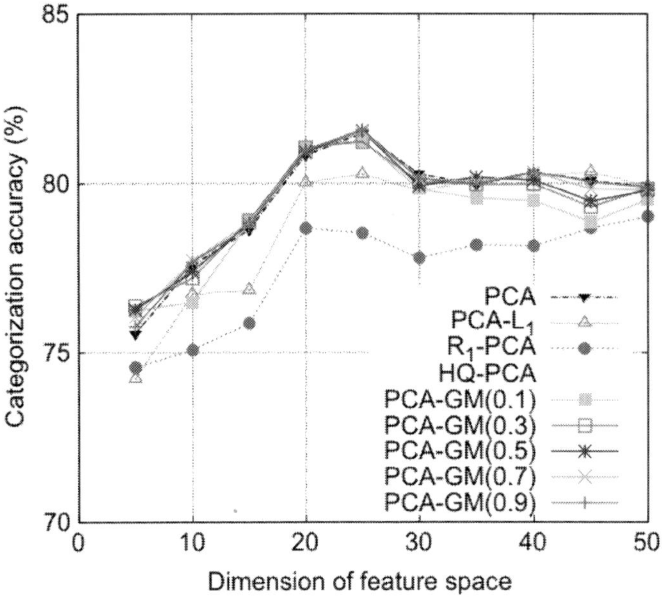
85
80
75
70
0
10
20
30
40
50
Categorization accuracy (%)
Dimension of feature space
PCA
PCA-L1
R1-PCA
HQ-PCA
PCA-GM(0.1)
PCA-GM(0.3)
PCA-GM(0.5)
PCA-GM(0.7)
PCA-GM(0.9)

c

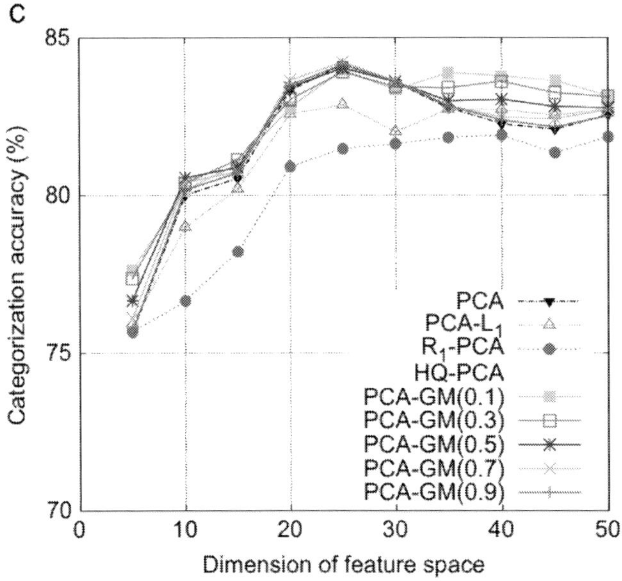
85
80
75
70
0
10
20
30
40
50
Categorization accuracy (%)
Dimension of feature space
PCA
PCA-L1
R1-PCA
HQ-PCA
PCA-GM(0.1)
PCA-GM(0.3)
PCA-GM(0.5)
PCA-GM(0.7)
PCA-GM(0.9)

d

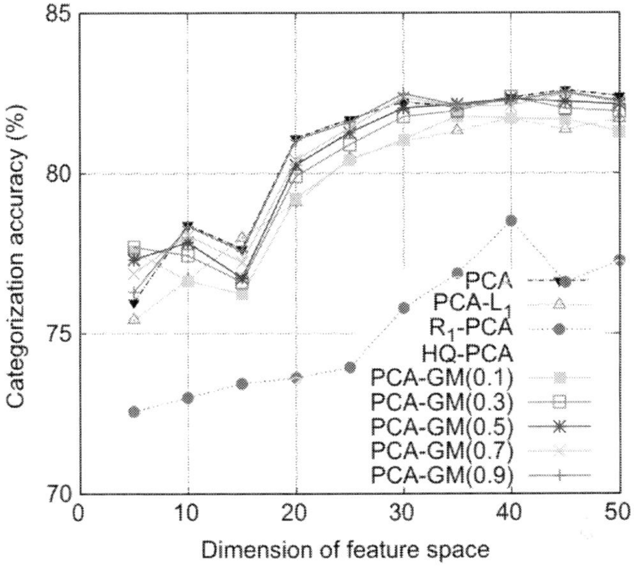
85
80
75
70
Categorization accuracy (%)
0
10
20
30
40
50
Dimension of feature space
PCA
PCA-L1
R1-PCA
HQ-PCA
PCA-GM(0.1)
PCA-GM(0.3)
PCA-GM(0.5)
PCA-GM(0.7)
PCA-GM(0.9)

e

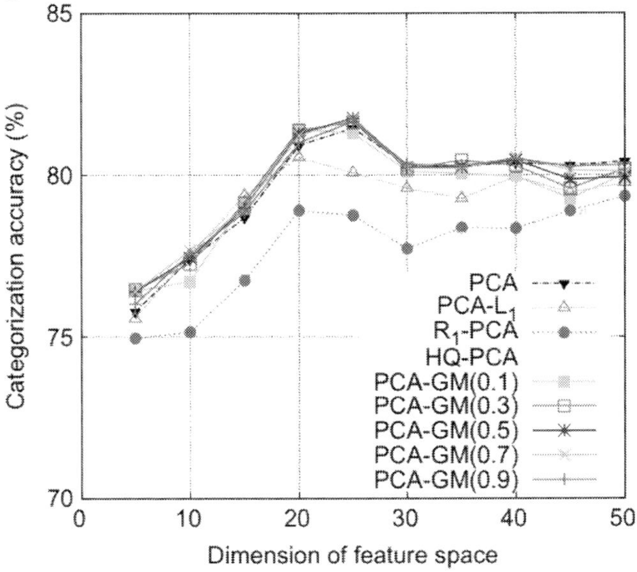
85
80
75
70
Categorization accuracy (%)
0
10
20
30
40
50
Dimension of feature space
PCA
PCA-L1
R1-PCA
HQ-PCA
PCA-GM(0.1)
PCA-GM(0.3)
PCA-GM(0.5)
PCA-GM(0.7)
PCA-GM(0.9)

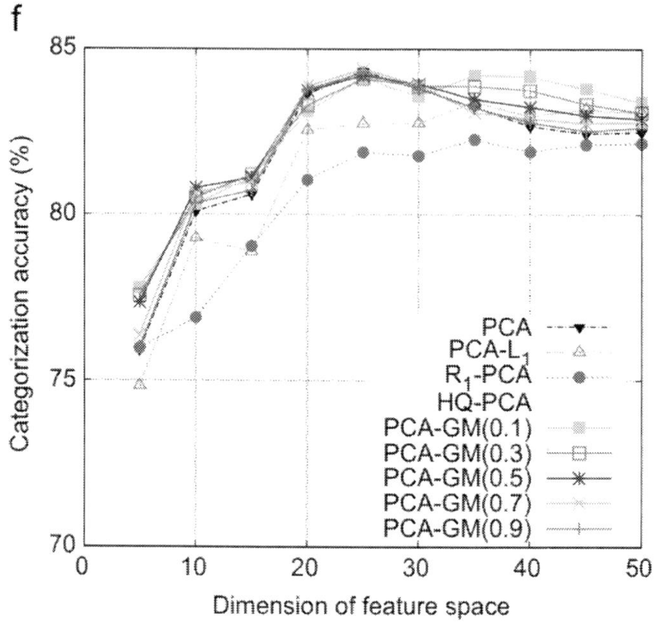

Figure 9. Categorization accuracy of different PCA methods for the training sets with different sizes of training images (n) and different numbers of training samples (N); (a) (48×48, 3375), (b) (48×48, 12,150), (c) (48×48, 24,300), (d) (64×64, 3375), (e) (64×64, 12,150), (c) (64×64, 24,300).

Together with the categorization accuracy, we measured number of iterations in PCA-GM and running time of the proposed method to obtain projection matrices from the above six training sets of the Small NORM data set. Table 2 shows the average numbers of iterations performed in the proposed method. From this table, we can find that PCA-GM converges in less than 50 iterations on average. Also, the average number of iterations decreases as the value of p increases from 0.1 to 0.9. This may have been resulted from the fact that the objective function of PCA-GM has many fluctuations when the value of p is close to zero whereas it is similar to one of the conventional PCA, which is quadratic, when the value of p is close to one. The overall running time of the proposed method described in Algorithm 2 varies depending on the number of iterations needed until a stop criterion is satisfied. Thus, we divided the overall running times by the average numbers of iterations performed in computing five projection matrices with respect to five categories for every combination of m, n, and N, which are summarized in Table 3 in the setting of p=0.1. From the other values of p, we could see the similar tendencies. The running times were measured on a 3.4 GHz Intel Xeon workstation with 12 cores using MATLAB. Each iteration in the algorithm consists of two processes, the approximation and the minimization. Compared to the approximation, the minimization requires much more computations. It corresponds to the weighted eigenvalue decomposition, which was implemented by applying the singular value decomposition (SVD) to the weighted data matrix instead of computing the weighted covariance matrix and applying the eigenvalue decomposition to it for efficiency. Thus, the running times reported in Table 3 can be regarded as the running time of the SVD approximately. Considering the average numbers of iterations shown in Table 2, it can be said that the proposed method is feasible enough until N=25,000 and n=5000 roughly.

Table 2. Average numbers of iterations needed in PCA-GM on small NORB data set

p=0.1	p=0.3	p=0.5	p=0.7	p=0.9
22.89	23.43	19.27	14.80	8.42

Table 3. Average running time in seconds per each iteration in PCA-GM with p=0.1 on small NORB data set

m	n=48×48			n=64×64		
	N=3375	N=12,150	N=24,300	N=3375	N=12,150	N=24,300
5	1.05	22.55	40.28	1.62	47.37	146.31
10	0.95	25.45	42.90	1.59	51.36	150.91
15	0.98	25.69	37.78	1.59	52.55	134.89
20	0.94	23.74	39.14	1.49	46.61	149.55
25	0.89	17.99	35.50	1.44	36.83	130.75
30	0.88	19.60	38.86	1.42	42.74	98.59
35	0.89	18.34	39.40	1.46	43.60	136.14
40	0.84	20.23	32.62	1.42	42.51	106.12
45	0.85	20.21	33.70	1.38	39.35	126.85
50	0.84	18.35	32.03	1.43	41.43	111.99

CONCLUSION AND DISCUSSION

We proposed a robust PCA using the generalized mean to mitigate the negative effect of outliers belonging to the training set. Considering the fact that the sample mean is prone to the outliers, a generalized sample mean was proposed based on the generalized mean as an alternative to the sample mean in the framework of the proposed method. The efficient iterative methods were also developed to solve the optimization problems formulated using the generalized mean. Experiments on the face reconstruction, clustering, and object categorization problems demonstrated that the proposed method performs better than or equal to the other robust PCAs depending on the problems tackled. We expect that the proposed methods can be used in various applications. For example, a trimmed average, which is one of the robust first-order statistics, was used in a scalable robust PCA method [29]. We think that the generalized sample mean can be an effective alternative to the trimmed average.

ACKNOWLEDGMENTS

This research was supported by Basic Science Research Program through the National Research Foundation of Korea (NRF) funded by the Ministry of Education (NRF-2013R1A1A2007461, NRF-2013R1A1A1006599).

REFERENCES

1. A. Jain, R. Duin, M. Jianchang, Statistical pattern recognition: a review, IEEE Trans. Pattern Anal. Mach. Intell. 22 (1) (2000) 4–37.
2. I. Jolliffe, Principal Component Analysis, 2nd edition, Springer-Verlag, New York, 2002.
3. M. Turk, A. Pentland, Eigenfaces for recognition, J. Cognit. Neurosci. 3 (1) (1991) 71–86.
4. P. Belhumeur, J. Hespanha, D. Kriegman, Eigenfaces vs. Fisherfaces: recognition using class specific linear projection, IEEE Trans. Pattern Anal. Mach. Intell. 19 (7) (1997) 711–720.
5. D.A. Ross, J. Lim, R.-S. Lin, M.-H. Yang, Incremental learning for robust visual tracking, Int. J. Comput. Vis. 77 (1–3) (2008) 125–141.
6. K.Y. Yeung, W.L. Ruzzo, Principal component analysis for clustering gene expression data, Bioinformatics 17 (9) (2001) 763–774.

7. C. Ding, X. He, K-means clustering via principal component analysis, in: Proceedings of the 21st International Conference on Machine Learning, ICML '04, 2004.
8. F. de la Torre, M.J. Black, A framework for robust subspace learning, Int. J. Comput. Vis. 54 (1–3) (2003) 117–142.
9. Q. Ke, T. Kanade, Robust L1 norm factorization in the presence of outliers and missing data by alternative convex programming, in: IEEE Computer Society Conference on Computer Vision and Pattern Recognition, 2005, vol. 1, 2005, pp. 739–746.
10. C. Ding, D. Zhou, X. He, H. Zha, R1-PCA: rotational invariant L1-norm principal component analysis for robust subspace factorization, in: Proceedings of the 23rd International Conference on Machine Learning, ICML '06, 2006, pp. 281– 288.
11. N. Kwak, Principal component analysis based on L1-norm maximization, IEEE Trans. Pattern Anal. Mach. Intell. 30 (9) (2008) 1672–1680.
12. R. He, B. Hu, X. Yuan, W.-S. Zheng, Principal component analysis based on nonparametric maximum entropy, Neurocomputing 73 (10–12) (2010) 1840–1852.
13. R. He, B.-G. Hu, W.-S. Zheng, X.-W. Kong, Robust principal component analysis based on maximum correntropy criterion, IEEE Trans. Image Process. 20 (6) (2011) 1485–1494.
14. N. Kwak, Principal component analysis by lp-norm maximization, IEEE Trans. Cybern. 44 (5) (2014) 594–609.
15. Z. Liang, S. Xia, Y. Zhou, L. Zhang, Y. Li, Feature extraction based on Lp-norm generalized principal component analysis, Pattern Recognit. Lett. 34 (9) (2013) 1037–1045.
16. J. Brooks, J. Dulá, E. Boone, A pure L1-norm principal component analysis, Comput. Stat. Data Anal. 61 (2013) 83–98.
17. A.Y. Ng, Feature selection, L1 vs. L2 regularization, and rotational invariance, in: Proceedings of the 21st International Conference on Machine Learning, ICML '04, 2004.
18. T.M. Cover, J.A. Thomas, Elements of Information Theory, John Wiley & Sons, New York, 1981.
19. W. Liu, P.P. Pokharel, J.C. Principe, Correntropy: properties and

applications in non-Gaussian signal processing, IEEE Trans. Signal Process. 55 (11) (2007) 5286–5298.

20. P. Bullen, Handbook of Means and Their Inequalities, 2nd edition, Springer, Netherlands, 2003.
21. E.J. Candès, X. Li, Y. Ma, J. Wright, Robust principal component analysis? J. ACM 58 (3) (2011) 11:1–11:37.
22. P.J. Huber, E.M. Ronchetti, Robust Statistics, 2nd edition, Wiley, New Jersey, 2009.
23. G.H. Golub, C.F.V. Loan, Matrix Computations, 3rd edition, Johns Hopkins, Baltimore, 1996.
24. J. Oh, N. Kwak, M. Lee, C.-H. Choi, Generalized mean for feature extraction in one-class classification problems, Pattern Recognit. 46 (12) (2013) 3328–3340.
25. C.M. Bishop, Pattern Recognition and Machine Learning, Springer-Verlag, New York, 2006.
26. D. Comaniciu, P. Meer, Mean shift: a robust approach toward feature space analysis, IEEE Trans. Pattern Anal. Mach. Intell. 24 (5) (2002) 603–619.
27. P. Phillips, H. Moon, S. Rizvi, P. Rauss, The FERET evaluation methodology for face-recognition algorithms, IEEE Trans. Pattern Anal. Mach. Intell. 22 (10) (2000) 1090–1104.
28. Y. LeCun, F.J. Huang, L. Bottou, Learning methods for generic object recognition with invariance to pose and lighting, in: Proceedings of the 2004 IEEE Computer Society Conference on Computer Vision and Pattern Recognition, vol. 2, 2004, pp. II-97–104.
29. S. Hauberg, A. Feragen, M. Black, Grassmann averages for scalable robust pca, in: Proceedings of the 2014 IEEE Conference on Computer Vision and Pattern Recognition, 2014, pp. 3810–3817.

CHAPTER 12

INTRODUCING COMPLEXITY CURTAILING TECHNIQUES FOR THE TOUR CONSTRUCTION HEURISTICS FOR THE TRAVELLING SALESPERSON PROBLEM

Ziauddin Ursani[1,2] and David W. Corne[1,2]

[1]School of Mathematical and Computer Sciences, Heriot Watt University, Edinburgh EH14 4AS, UK

[2]Route Monkey Ltd., Livingston EH54 5DW, UK

ABSTRACT

In this paper, complexity curtailing techniques are introduced to create faster version of insertion heuristics, that is, cheapest insertion heuristic (CIH) and largest insertion heuristic (LIH), effectively reducing their complexities from $O(n^3)$ to $O(n^2)$ with no significant effect on quality of solution. This paper also examines relatively not very known heuristic concept of max difference and shows that it can be culminated into a full-fledged max difference insertion heuristic (MDIH) by defining its missing steps. Further to this the paper extends the complexity curtailing techniques to MDIH to create its faster version. The resultant heuristic, that is, fast max difference insertion heuristic (FMDIH), outperforms the "farthest insertion" heuristic (FIH) across a wide spectrum of popular datasets with statistical significance, even though both the heuristics have the same worst case complexity of $O(n2\ )$. It should be noted that FIH is considered best among lowest order complexity heuristics. The complexity curtailing techniques presented here open up the new area of research for their possible extension to other heuristics.

INTRODUCTION

The Traveling Salesman Problem (TSP) is one of the most studied problems in the scientific literature and sometimes is referred to as a mother of all combinatorial optimization problems. It continues to be a testing ground for the development of combinatorial optimization methods, while having numerous practical applications in diverse areas, including logistics, genetics, manufacturing, telecommunications, and neuroscience [1]. Examples of real-world application domains with problems that can be naturally formulated as the TSP include VLSI design, vehicle routing, data clustering, and job-shop scheduling [2]. The earliest reference to the TSP can be found in the 1832 German handbook for travelling salesmen [1]. The problem consists of finding the shortest possible tour of a set of n cities, such that the tour starts from and ends at the same city and visits each of the remaining cities precisely once. The problem is simple to state but has proved to be intractable and is included among the seven "millennium prize problems" described by the Clay Mathematical Institute, carrying a prize of one million US dollars for discovery of a polynomial time solution method.

Meanwhile, research and practice in the TSP have focused on heuristic methods that yield fast approximate solutions. These heuristic methods tend to cluster into three main groups:

- tour construction heuristics,
- tour improvement heuristics, and
- composite heuristics (which combine elements of both tour construction and tour improvement).

A tour construction heuristic constructs the tour from scratch, beginning with one city and iteratively expanding the subtour by one city at a time. In contrast, a tour improvement heuristic begins with a complete tour and makes one or more rearrangements in an attempt to improve it. A vast space of "composite" heuristics obscures the distinction between these two categories by using elements of both; for example, in the highly successful Lin-Kernighan heuristic [3], the context is that of iterated tour improvement; however the improvement process consists of repeated construction of full solutions from partial solutions. The concord software code [4], which has solved most problems in TSPLIB to optimality, uses the Lin Kernighan heuristic to find out near optimal tours and then applies various mathematical programming methodologies to achieve optimality.

With no general polynomial time solution method yet available, the quest for the development of new heuristics for the TSP remains active. Tour construction heuristics have played an important role in this quest, in particular because they can form components of a wide range of approaches. For example, they can be used to construct the starting solutions for tour improvement heuristics [5], and they can provide rough estimates of the cost of optimal solutions. In turn, they therefore provide interesting analytical grounds for the study of upper bounds on solution quality [6] and provide material for empirical studies [7, 8] that attempt to understand how optimal solutions differ from solutions that arise from tour construction heuristics. Such studies lead to better understanding of the structure of optimal solutions of the TSP, a problem that has now remained centre of our intellectual curiosity for centuries.

The basic procedure of tour construction heuristics can be summarized as follows:

- Establishment of the initial small subtour (subtour establishment rule).
- Selection of a city not present in the current subtour (selection rule).
- Expansion of the current subtour to include the selected city (expansion rule).
- Iterative application of steps (2) and (3) until a complete tour is obtained.

In the repeated application of steps (2) and (3), the subtour is "successively augmented" with the insertion of a new city; this is why tour construction heuristics are sometimes also called "successive augmentation" heuristics (see, e.g., [9]). Different tour construction heuristics are characterized by the specific methods they choose for each of the three steps: the initial subtour establishment, selection, and expansion rules. Expansion rules can be grouped into two main types: insertion and addition. An insertion-based expansion rule chooses where in the permutation to place the new city on the basis of the cost of the resulting subtour, whereas an addition-based expansion rule bases this decision on next-hop distance. It has been reported that insertion heuristics generally perform better than addition heuristics (see, e.g., [8]); we have therefore chosen to focus on insertion-based expansion heuristics in this paper.

Three insertion heuristics are particularly prominent in the literature, namely, nearest, farthest, and cheapest [10]. We consider these three heuristics in the remainder of this work, along with a further two that are less well known. The first of these two is the "largest-cost insertion heuristic" (LIH) [11], which sits naturally alongside the aforementioned three, despite being rarely considered in the literature. The second is the "max difference insertion heuristic" (MDIH). Tunnel and Heath [12] presented "max difference" as a concept that could be attached to any other heuristic; however we argue that, with appropriate extensions that we later describe (resulting in MDIH), it is more appropriately seen as a tour construction heuristic in itself, and in fact we demonstrate that it is a particularly effective one.

The insertion heuristics under study can be divided into two groups, that is, distance based insertion heuristics (DBIH) and cost based insertion heuristics (CBIH). The DBIH has city selection rule based on distance. This group includes nearest (NIH) and farthest insertion heuristics (FIH). The CBIH has city selection rule based on the cost of insertion (see (5)). This group includes cheapest (CIH), largest (LIH), and max difference insertion heuristics (MDIH). The difference in selection rule brings about difference in worst case complexity of heuristics with DBIH having worst case complexity of $O(n^2)$ and CBIH having complexity of $O(n^3)$ (see Section 4). This paper proposes techniques to curtail the complexity of CBIH to (n^2), effectively creating faster versions of CBIH, that is, FCIH, FLIH, and FMDIH, while addition of "F" denotes the word fast.

Previously it has been reported that farthest insertion heuristic (FIH) generally performs best among the community of insertion heuristics of (n^2) or lower time complexity (see, e.g., [13]). However we will show that the performance of FMDIH is consistently better than FIH on a wide spectrum of popular datasets even though the complexity of FMDIH is no greater than that of FIH.

The remainder of this paper is structured as follows. In Section 2, we describe four of the five insertion heuristics of interest: cheapest, largest, nearest, and farthest insertion. In Section 3 we describe the previously overlooked "max difference" concept and build on that to describe the "max difference insertion" heuristic. In Section 4, we discuss design and complexity issues for all five of the heuristics of interest, and in Section 5 we then describe our complexity curtailing techniques to produce accelerated and approximated variants of cost based insertion heuristics. Section 6 provides summaries of empirical results, and we conclude with a statement of our main findings in Section 7. In Appendix, we then show the empirical results for the farthest and max difference insertion heuristics in finer detail.

DESCRIPTION OF TOUR CONSTRUCTION HEURISTICS

Below, we explain the design details of four of the five main heuristics considered in this paper. Our elaboration of the details follows the four-part structure given in Section 1 for tour construction heuristics. Being insertion heuristics, the most important elements are the city selection and subtour expansion methods. The city selection method is also salient for another reason, which is the fact that the name of a tour construction heuristic (e.g., cheapest and farthest) tends to reflect the nature of this rule. We come back to this fact in Section 3, when we speculate about why the "max difference" heuristic has been largely overlooked. The simple "initial subtour establishment" rule is also described for completeness and to facilitate replication of our experiments. Meanwhile, it is worth noting an interesting aspect of the city selection rule.

Initial Subtour Establishment

This is the first step in any tour construction heuristic. In each of the standard forms of the cheapest, largest, nearest, and farthest insertion heuristics, this initial "subtour" is simply a single city chosen uniformly at random.

City Selection

Momentarily referring to the subtour expansion rule (in Section 2.3), we note that two of the heuristics of interest are based on cost, and two are based on distance. Where subtour expansion is based on cost, the city selection rule is as follows. Let τ be the set of cities in the subtour and let τ' be the set of cities not in the subtour, while city $k \in \tau'$ and cost(τ, k) is the expansion cost of subtour τ by insertion of city k. Then, the city k is selected such that

for CIH

$$\text{cost}(\tau, k) = \min\left\{\forall j \in \tau' \;\; (\text{cost}(\tau, j))\right\} \tag{1}$$

and for LIH

$$\text{cost}(\tau, k) = \max\left\{\forall j \in \tau' \;\; (\text{cost}(\tau, j))\right\}. \tag{2}$$

When subtour expansion is instead based on distance, the city selection rule is different. Now, let (h, k) be the distance between cities k and h; the city k is selected such that

for NIH

$$d(h, k) = \min\left\{\forall j \in \tau', \;\; \forall i \in \tau \;\; (d(i, j))\right\} \tag{3}$$

and for FIH

$$d(h, k) = \max\left\{\forall j \in \tau', \;\; \forall i \in \tau \;\; (d(i, j))\right\}. \tag{4}$$

Subtour Expansion

As indicated in the introduction, subtour expansion heuristics can be based around either insertion or addition. We are only concerned with insertion based variants in this paper and so will confine our description accordingly. Let $k \in \tau'$ and (i, i')$\in \tau$, such that i and i' are a pair of consecutive cities in τ. Let (i, k) be the distance between the cities i and k. The subtour is expanded by inserting the selected city k in the subtour τ such that

$$\text{cost}(\tau, k) = \min\left[\forall\left(i,i'\right) \in \tau \ \left\{d(i,k) + d\left(i',k\right) - d\left(i,i'\right)\right\}\right]. \tag{5}$$

This step is applicable to each of CIH, LIH, FIH, and NIH. From this point onwards, we refer to the edge (i, i') related to cost(τ, k) as the expansion cost edge. It should be noted that the "largest-cost insertion" heuristic is not published yet, however, an unpublished paper about it is present on the internet [10]. We include it in our experiments, noting that it is the logical counterpoint to CIH, in the same way that FIH is related to NIH. It therefore completes a set of four heuristics of which two are based on minimum and maximum distance ((3)-(4)) while two are based on minimum and maximum expansion cost ((1)-(2)). We later find that LIH outperforms both CIH and NIH, though apparently missing from the literature.

THE MAX DIFFERENCE INSERTION HEURISTIC (MDIH)

The idea behind the MDIH was introduced in a Master's thesis [12] a quarter of a century ago and no research paper about this could be found in literature. The "max difference" concept was presented by Tunnel and Heath as a way to engineer new variants of existing heuristics (they applied it to two versions of CIH and also to Stewart's Algorithm [14]). However, since (as we will see) the concept relates directly to how the city selection step is done, from which the names of tour construction heuristics seem invariably to be derived, we would argue that it merits reinvention as a fully fledged tour construction heuristic. Its performance in the form of MDIH, as detailed in the next section, certainly supports this view. Meanwhile we speculate that Tunnel and Heath may not have put it forward as a standalone new heuristic, based on the belief that it was the "insertion" (subtour expansion) rather than the "max difference" (city selection) that was salient in such categorization; in turn, this seems to have led to its being unnoticed, despite its comparatively strong performance among similar heuristics.

We now explain the key "city selection" step for MDIH. First, we need some helper definitions. The first of these is the function "nth minimum," denoted as min{∀V ∈U(V)}. The meaning of "nth minimum" is that if all the values V in set U are sorted from minimum to maximum then this function represents the nth value in that list. Now, the "nth expansion cost" of a subtour by the insertion of any city not in that subtour is equal to the function min$n${∀V ∈U(V)}, if the set U is the set of all the expansion costs of that subtour on all of its edges caused by the insertion of that city. Now let k∈τ' and (i, i')∈τ, such that i and i' are a pair of consecutive cities in τ. Let (i, k) be the distance between the cities i and k. Let cost(τ, k) represent nth expansion cost of the tour τ by the insertion of city k and then by the definition

$$\text{cost}_n(\tau,k)$$

$$=\min_n\left[\forall\left(i,i'\right)\in\tau\ \left\{d(i,k)+d\left(i',k\right)-d\left(i,i'\right)\right\}\right]. \tag{6}$$

It should be noted that, apart from here, the exclusive use of above function is made in Sections 5.2 and 5.3. From this point onwards the edge (i, i') related to cost$n$(τ, k) is referred to as the nth expansion cost edge. In the city selection rule of MDIH the city k is selected such that the cost difference D(k) between its expansion cost and 2nd expansion cost is maximum among all the cities not in the tour; that is,

$$D(k)=\max\left[\forall k\in\tau'\ \left\{\text{cost}_2(\tau,k)-\text{cost}(\tau,k)\right\}\right]. \tag{7}$$

It should be noted that there is no difference between cost1(τ, k) and cost(τ, k). The 3rd step (subtour expansion) follows the same rule as represented by (5).

Finally to present MDIH as complete tour construction heuristic in itself, we also explore five variants of the subtour establishment rule (step (1)). Unlike the other insertion heuristics considered in this paper, the city selection step of MDIH requires an initial subtour containing at least three cities. We therefore test five subtours establishment rules:

- a subtour of three randomly chosen cities;
- a subtour first formed by two randomly chosen cities and then the third city is chosen based on cheapest expansion;

- a subtour first formed by two randomly chosen cities and then the third city is chosen based on largest-cost expansion;
- a subtour initially formed by one randomly chosen city and the next two are chosen iteratively based on cheapest expansion;
- a subtour first formed by one randomly chosen city and then the next two are iteratively chosen on the basis of largest-cost expansion. Associated experiments and results are discussed later in Section 6.

DESIGN AND COMPLEXITY ASPECTS OF TOUR CONSTRUCTION HEURISTICS

Design of any heuristic consists of the design of data structures and algorithms for its efficient implementation. Our implementation centers around two key data structures: a doubly circular linked list representing the set of cities not in the subtour and a singly circular linked list representing the subtour. For set of cities not in the subtour, the doubly circular linked list allows for a very simple and fast city deletion operation. For the set of cities representing the subtour a, single circular linked list is sufficient since only an insertion operation is needed here. Initially, a doubly circular linked list is prepared representing the set of cities not in the subtour. From this list, cities are deleted one by one and correspondingly inserted in the single circular linked list representing the subtour of cities at each step. The procedure continues until the doubly circular linked list becomes empty.

The design of step (2) (city selection) of DBIH, that is, the nearest and farthest insertion heuristics ((3)-(4)), consists of finding the nearest neighbor city present in the subtour for each of the cities not in the subtour. To reduce computation costs, the nearest neighbour city in the subtour for each city not in the subtour is recorded in each iteration, and in the next iteration the distance of only the recently added city is compared with the distance of the nearest neighbor from the previous iteration, thus updating the nearest neighbor information for the current iteration. Therefore, FIH and NIH can be implemented within time complexity of $O(n^2)$.

In the case of CBIH, that is, the cheapest and largest insertion heuristics, the design of the city selection step ((1) and (2)) consists of finding the expansion cost for insertion into the subtour of each city not in the subtour. This implies that the expansion cost needs to be computed for each city not in the subtour; in turn, computation of the expansion requires visits to each and every edge of the subtour. To make this efficient, information should be preserved between iterations, and in each new iteration the expansion cost of only newly added edges is computed and compared with the expansion costs of the previous iteration to update the expansion cost information. However this shortcut may not be possible for all the cities not in the subtour. This is because some of the cities may have expansion costs in the previous iteration at an edge which is broken in the current iteration. This means the information about the expansion cost of those cities is no longer valid since the relevant edge for the expansion does not exist. For such cities, we need to compute the expansion costs for newly formed edges and if this expansion cost is less than or equal to their expansion cost in previous iterations, then it means their new expansion cost has been found on the newly formed edges. However, if not, we are forced to recompute the expansion cost of these cities on all edges of the current subtour to find out their new expansion cost. Therefore worst case scenario complexity of CIH and LIH is (n^3). The average complexity of CIH is reported as of the order $(n^2 \log n)$ [14].

To design the city selection step for MDIH, for each city not in the subtour both the expansion cost and the 2nd expansion cost need to be computed (7). However, both of these quantities can be computed simultaneously in a single scan of the subtour. To implement the step efficiently, these quantities should be preserved between iterations for each city not in the subtour, and in the new iteration only the expansion cost and 2nd expansion cost of the recently added edges are computed and compared with those of the previous iteration to update them. However, just as was the case for CIH and LIH, this shortcut is not always possible; for some cities the expansion costs at all edges of the subtour need to be computed. Therefore the worst case scenario complexity of MDIH is same as that of cheapest and largest insertion heuristics, that is, (n^3).

COMPLEXITY CURTAILING TECHNIQUES FOR COST BASED INSERTION HEURISTICS

In this section, complexity curtailing techniques are introduced to create fast and approximate variants of CIH, LIH, and MDIH.

Design of Fast Cheapest Insertion Heuristic (FCIH)

We can devise a new but faster heuristic based on CIH on the basis of simple geometrical principle. Consider the subtour "$abcdefa$" in Figure 1.

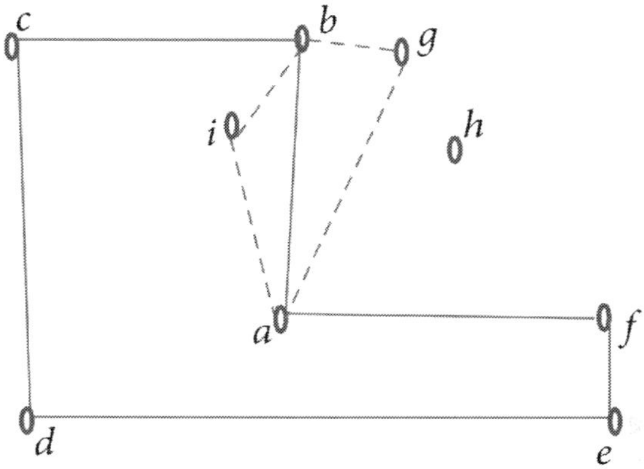

Figure 1: Subtour "$abcdefa$" showing principle of fast CIH.

In Figure 1 it can be seen that g, h, i are the cities yet to be inserted. For each of these cities, the cheapest insertion edge is ab. The final shape of the tour depends on the order in which these three edges are inserted. If city i is inserted first, then edge ab will be deleted and the edges ia and ib will be added. In the new subtour $aibcdefa$ it is now a genuine possibility that the cheapest insertion for city h becomes edge af, since edge ab is no longer present. In this case, we will need to traverse the entire current subtour $aibcdefa$ to determine the new cheapest insertion for h. However, if city g had been inserted first instead of city i then edge ab is deleted and new edges ga and gb are added. In this case most likely situation becomes that h is inserted in the newly added edge ga, requiring only inspection of the newly added edges ga and gb.

In interim summary, if we encounter the scenario in Figure 1 during the construction of a tour using CIH, then, for the sake of computational efficiency, we would hope that city i is not chosen as the first of the three to be inserted. However, some reflection will reveal that the position of city i in this scenario is unlikely in the first place. It is well known that CIH grows tours "outwardly." This means that the cities not in the subtour lie outside the periphery of the subtour, rather than inside it, such as we see with city i in Figure 1. By exploiting this property, we can propose a variant of CIH that only inspects the recently added edges in each iteration, even for the cities whose earlier cheapest edge is broken. Though no longer guaranteeing the "cheapest" insertion, intuition suggests that this would lead to a favorable tradeoff, with at most minor loss of quality set against significant benefit in speed. In this "fast CIH" heuristic, the total number of computations of expansion costs (now independent of details of the dataset) is given below:

$$N_c = 2 \times \{(n-1) + (n-2) + (n-3) + \cdots + 3 + 2 + 1\}$$

$$= n(n-1). \quad (8)$$

Meanwhile, since one expansion cost involves traversal of 3 edges, the total number of edges traversed for computation of expansion costs is given by

$$N_e = 3n(n-1). \quad (9)$$

The time complexity of "fast CIH" is therefore (n^2), which is below the average complexity of CIH. Note that fast CIH runs in the order (n^2) on any dataset (with best case, worst case, and average case scenarios all equivalent). A full procedural summary of this new "fast cheapest insertion" heuristic (FCIH) is given in Procedure 1.

Procedure FCIH starts
Form circular linked list of cities τ' not in the subtour
Form empty Linked list τ
Choose an arbitrary city $p \in \tau'$
Delete p from the list τ'
Insert p in the empty list τ
Initialize expansion costs of subtour τ by insertion w.r.t. all h in τ' to very high value
while τ' is not empty *do begin*
    do begin (Visit all h in τ')
        Let
        $\delta[1]$ = Cost of insertion of city h on 1st newly formed edge
        $\delta[2]$ = Cost of insertion of city h on 2nd newly formed edge
        $\delta[3] = \text{cost}(\tau, h)$

$\delta[3] = \text{cost}(\tau, h)$
If expansion cost edge is broken in last iteration
then $\text{cost}(\tau, h) = \min[\forall r \in \{1, 2\}\ (\delta[r])]$
else $\text{cost}(\tau, h) = \min[\forall r \in \{1, 2, 3\}\ (\delta[r])]$
end if
while all h in τ' not visited
Choose city q in τ' such that $\text{cost}(\tau, q) = \min\{\forall h \in T\ (\text{cost}(\tau, h))\}$
Delete q from the list τ'
Insert q in τ on the edge connected to $\text{cost}(\tau, q)$
end while
Procedure FCIH ends

Procedure 1: Fast cheapest insertion heuristic (FCIH).

The first five lines of "Procedure 1" establish the initial tour. Expansion of this initial tour is then implemented in the remainder of the procedure. The inner loop in the remainder is used to compute the expansion cost cost(τ, h) for each city h (the definition of this cost is given in (5)). However, a short cut is used to compute this cost, by computing the expansion cost on only two newly formed edges. There are two potential scenarios associated with this shortcut: in the first scenario, the edge related to the expansion cost of this city in the previous iteration is broken, and in the second scenario it has survived. In the first scenario, the new expansion cost of the city h in the current iteration is assumed to be the minimum of the two expansion costs of city h computed on the two newly formed edges; in the second scenario, the difference is that we also consider the expansion costs of city h computed in the previous iteration. The condition "*while* all h in τ' not visited" ensures that the cost of all uninserted cities is computed. The outer loop chooses the city for insertion that has the lowest expansion cost.

Design of Fast Largest Insertion Heuristic (FLIH)

In contrast to CIH, tours constructed by LIH do not grow "outwardly," and the situation in Figure 1, where one of the currently unvisited cities lies inside the periphery of the current subtour, is a common occurrence. We cannot therefore design a faster version of LIH on the same basis used for FCIH. However, in the case of LIH a different approximate assumption can be made which may help avoid computing expansion costs on all edges of the current subtour. Referring again to Figure 1:

if city i is inserted in the subtour first, the new subtour $aibcdefa$ is formed, breaking the edge ab as a result. Now in this case, as discussed earlier, city h might not make its cheapest insertion with newly formed edges ai and ib. However, there is a genuine possibility that city h makes its 2nd cheapest insertion with ai or ib. This leads us to the idea of a fast version of LIH in which we approximate, rather than guarantee, the cheapest insertion, while scanning at most four edges in each iteration of the construction process. In detail, in each iteration we only need to keep a record of the 2nd expansion cost in addition to the (1st) expansion cost for each city not in the subtour. In turn, we find the 2nd expansion cost of a city by scanning at most only four edges: the two newly formed edges, along with the expansion cost edge and 2nd expansion cost edge of the previous iteration. If either of the latter two were broken in the previous iteration, then we can simply use the remaining three edges to (approximately) update the expansion cost and 2nd expansion cost in the new iteration. Again, the resulting heuristic, FLIH, has complexity (n^2), independently of details of the dataset. A procedural summary of fast largest insertion heuristic (FLIH) is given in Procedure 2.

Procedure FLIH starts
Form circular linked list of cities τ' not in the subtour
Form empty Linked list τ
Choose an arbitrary city $p \in \tau'$
Delete p from the list τ'
Insert p in the empty list τ
Initialize expansion costs and 2nd expansion costs w.r.t. all h in τ' to very high value
while τ' is not empty do begin
    do begin (Visit all h in τ')
        Let
        $\delta[1]$ = Cost of insertion of city h on 1st newly formed edge
        $\delta[2]$ = Cost of insertion of city h on 2nd newly formed edge
        $\delta[3] = \text{cost}(\tau, h)$

```
            δ[4] = cost_2(τ, h)
            If          expansion cost edge is broken in last iteration
            then        ∀n ∈ {1,2} (cost_n(τ,h) = min_n[∀r ∈ {1,2,4} (δ[r])])
            else if     2nd expansion cost edge is broken in last iteration
            then        ∀n ∈ {1,2} (cost_n(τ,h) = min_n[∀r ∈ {1,2,3} (δ[r])])
            else        ∀n ∈ {1,2} (cost_n(τ,h) = min_n[∀r ∈ {1,2,3,4} (δ[r])])
            end if
      while all h in τ' not visited
      Choose city q in τ' such that cost(τ,q) = max{∀h ∈ T (cost(τ,h))}
      Delete q from the list τ'
      Insert q in τ on the edge connected to cost(τ,q)
end while
Procedure FLIH ends
```

Procedure 2: Fast largest insertion heuristic (FLIH).

In common with Procedure 1, the first five lines of "Procedure 2" establish the initial subtour, while the remainder concern its expansion. In the inner loop, the expansion cost cost(τ, h) and second expansion cost cost2(τ, h) are calculated for each city h (see (5) and (6), resp.). The costs are calculated on the basis of two newly formed edges, and the detail of the procedure concerns three potential scenarios that may arise. In the first scenario, the edge related to the expansion cost of city h in the previous iteration is broken. In the second, the edge related to the second expansion cost of city h is broken; in the third scenario, none of the expansion edges is broken. In the first scenario the new expansion cost and second expansion cost of city h are taken to be the two minimum expansion costs of three quantities: the two expansion costs of city h computed on the two newly formed edges and its second expansion cost in the last iteration. In the second scenario, the three quantities are now the two expansions costs computed on the two newly formed edges (as before) and its expansion cost computed in the previous iteration. Finally, in the third scenario, the two costs are taken as the smallest two of four quantities: the two expansion costs of city h computed on the two newly formed edges and its expansion cost and second expansion cost computed in the earlier iteration. The condition "*while* all h in τ' not visited" ensures that cost of all uninserted cities is computed. The outer loop chooses the city for insertion that has largest expansion cost.

Design of Fast Max Difference Insertion Heuristic (FMDIH)

Faster design of MDIH is possible on a similar basis to that used in faster design of FLIH. A key difference between LIH and FLIH is that, in each iteration, LIH calculates only the expansion costs for the unvisited cities, while FLIH (approximately) computes both the expansion costs and the 2nd expansion costs. By going "one step further" in this sense, FLIH facilitates much faster update of this information, while sacrificing some exactness. To extend the same principle to fast MDIH, we will go "one step further" in the computations required for the city selection rule in MDIH. In MDIH, city selection requires both the expansion cost and the 2nd expansion cost for each city not in the subtour (7). Therefore, we can design a fast MDIH that additionally computes and records the 3rd expansion cost of each city not in the subtour. This means we need to record at most five edges per iteration: the two newly added, plus the expansion, 2nd expansion, and 3rd expansion cost edges from the previous iteration. As with FLIH, if one of the latter three was broken in the current iteration, we will simply rely on the others to provide approximations. A detailed procedural summary of the fast max difference insertion heuristic (FMDIH) is given in Procedure 3.

Procedure FMDIH starts
Form circular linked list of cities τ' not in the subtour
Form empty Linked list τ
Setup initial subtour of three cities $(p_1, p_2, p_3) \in \tau'$
Delete (p_1, p_2, p_3) from the list τ'
Insert (p_1, p_2, p_3) in the empty list τ
Initialize expansion costs, 2nd expansion costs and 3rd expansion costs w.r.t. all h in τ' to very high value
while τ' is not empty do begin
    do begin (Visit all h in τ')
        Let
        $\delta[1]$ = Cost of insertion of city h on 1st newly formed edge
        $\delta[2]$ = Cost of insertion of city h on 2nd newly formed edge
        $\delta[3] = \text{cost}(\tau, h)$
        $\delta[4] = \text{cost}_2(\tau, h)$

$\delta[5] = \text{cost}_3(\tau, h)$
If expansion cost edge is broken in last iteration
then $\forall n \in \{1,2,3\}\ (\text{cost}_n(\tau, h) = \min_n[\forall r \in \{1,2,4,5\}\ (\delta[r])])$
else if 2nd expansion cost edge is broken in last iteration
then $\forall n \in \{1,2,3\}\ (\text{cost}_n(\tau, h) = \min_n[\forall r \in \{1,2,3,5\}\ (\delta[r])])$
else if 3rd expansion cost edge is broken in last iteration
then $\forall n \in \{1,2,3\}\ (\text{cost}_n(\tau, h) = \min_n[\forall r \in \{1,2,3,4\}\ (\delta[r])])$
else $\forall n \in \{1,2,3\}\ (\text{cost}_n(\tau, h) = \min_n[\forall r \in \{1,2,3,4,5\}\ (\delta[r])])$
end if
while all h in τ' not visited
Choose city q in τ' such that cost difference $D(q) = \max[\forall h \in T\{\text{cost}(\tau, h) - \text{cost}_2(\tau, h)\}]$
Delete q from the list τ'
Insert q in τ on the edge connected to $\text{cost}(\tau, q)$
end while

Procedure 3: Fast max difference insertion heuristic (FMDIH).

In "Procedure 3", again, the initial subtour is computed by the first five lines, and the remainder expands it. The inner loop of the remainder is used to compute the expansion cost cost(τ, h), the second expansion cost $\text{cost}_2(\tau, h)$, and the third expansion cost cost3(τ, h) for each city h. The definitions of these costs are given in (5) and (6). The details of the inner loop achieve a shortcut in computing these costs, which involves four scenarios. In the first scenario, the edge related to the expansion cost of city h in the previous iteration is broken. In the second scenario, the edge related to the second expansion cost of city h is broken; in the third scenario, the edge related to the third expansion cost of city h is broken, and in the fourth scenario none of the three expansion edges is broken. In the first scenario the expansion cost, second expansion cost, and third expansion cost are assumed to be the smallest three of four quantities: the two expansion costs of city h computed on the two newly formed edges and its second and third expansion costs in the last iteration. In the second scenario they are taken as the smallest three of a slightly changed set of four quantities: the two expansion costs of city h computed on the two newly formed edges and its expansion cost and third expansion cost computed in the earlier iteration. In the third scenario, the four quantities are now the two expansion costs of city h computed on the two newly formed edges and its expansion cost and second expansion cost from the earlier iteration. In the fourth scenario, first, second, and third expansion costs are taken as minimum costs of five quantities, that is, two expansion costs on two newly formed edges and first, second, and third expansion costs of the previous iteration. The condition "*while* all h in τ' not visited" ensures that cost of all uninserted cities is computed. The outer loop chooses the city for insertion that has maximum cost difference (7).

EXPERIMENTS AND RESULTS

This section consists of four subsections. In Section 6.1 we report experiments that evaluate each of the five types of initial subtour establishment rules described in Section 3 allowing us to "configure" and complete the new MDIH. Later in Section 6.2 performance of MDIH is compared with all baseline heuristics considered in this paper, that is, NIH, FIH, CIH, and LIH. In Section 6.3 the cost based insertion heuristics, that is, CIH, LIH, and MDIH, are compared with their faster versions FCIH, FLIH, and FMDIH. Finally in Section 6.4 the focus of our evaluation is to investigate the potential for FMDIH as a new $(n2\ )$ heuristic for fast approximate solution of TSPs, by comparing it with the current best-regarded $O(n2\ )$ heuristic, that is, FIH.

Experiments are conducted on 109 datasets, involving from 14 to 15112 cities from a popular test bed [14]. For each heuristic and problem instance, 30 simulations are run independently, and we record the best, worst, and average solutions, the standard deviation in solution quality, and the average execution time. For the majority of experiments we provide summary statistics in this section (volume of results data precludes otherwise); however a full description of results is provided in an appendix for the FMDIH and FIH tests. The hardware used in the experiments was an Intel Core i3-2348 M CPU @ 2.30 GHz, 8.0 GB Memory, with 64 bit operating system. The heuristics were coded in Microsoft Visual Studio C/C++.

Experiments to Finalize Initial Tour Establishment Step of Max Difference Insertion Heuristic (MDIH)

First of all, in Table 1 we can see a summary of the results for MDIH with the five different strategies for generating the initial subtour. The heuristics named MDIH-1 to MDIH-5 represent MDIH using methods (i) to (v) as described in Section 3. Each column in Table 1 refers to a certain statistical value in relation to deviation from the known optimal solution. The mathematical description of each of those values is given below, where C_1 to C_5 in (10)–(14), respectively, represent the quantities in columns 1–5 in Table 1:

$$C_1 = \text{avg}_{j=1}^{j=N}\left\{\min\left(\forall i \in S\ \left(\gamma_i^j\right)\right)\right\}, \quad (10)$$

$$C_2 = \text{avg}_{j=1}^{j=N}\left\{\max\left(\forall i \in S\ \left(\gamma_i^j\right)\right)\right\}, \quad (11)$$

$$C_3 = \text{avg}_{j=1}^{j=N}\left\{\text{avg}_{i=1}^{i=S}\left(\gamma_i^j\right)\right\}, \quad (12)$$

$$C_4 = \text{avg}_{j=1}^{j=N}\left[\text{avg}_{k=1}^{k=S}\left\{\gamma_k^j - \text{avg}_{i=1}^{i=S}\left(\gamma_i^j\right)\right\}\right], \quad (13)$$

$$C_5 = \text{avg}_{j=1}^{j=N}\left\{\text{avg}_{i=1}^{i=S}\left(t_i^j\right)\right\}, \quad (14)$$

where

$$\gamma_i^j = ((v_i^j - v_0^j)/v_0^j) \times 100,$$

$$\text{avg}_{i=1}^{i=S}(\gamma_i^j) = (\textstyle\sum_{i=1}^{i=S} \gamma_i^j)/S,$$

N means number of datasets which equals 109,

S means number of simulations which equals 30,

V^j_i means cost of solution of a simulation i on a dataset j,

V^j_0 means cost of optimal solution of a dataset j.

Table 1: Results of MDIH with five different initial subtour establishment rules

Name	Average of best solution % age	Average of worst solution % age	Average of average solution % age	Average of standard deviation % age	Average of execution time seconds
(0)	(1)	(2)	(3)	(4)	(5)
MDIH-1	4.56	12.12	7.37	1.78	0.45
MDIH-2	4.53	11.28	6.75	1.50	0.45
MDIH-3	4.47	9.09	6.50	1.12	0.46
MDIH-4	4.56	9.65	6.71	1.29	0.48
MDIH-5	4.64	8.59	6.34	0.99	0.46

In Table 1, it can be seen that MDIH-5 has the best average solution quality. Its average quality for worst solution is also among the best, while its standard deviation is also the best of the five. Only on one criterion, average of best solution, its performance is worst among all. However, the difference between the best and worst values for this statistic is only 0.17%, and we propose that it can be ignored in favour of choosing method (v) as the appropriate initial subtour establishment rule for MDIH and FMDIH.

Experiments to Compare Performance of All Baseline Heuristics NIH, CIH, LIH, FIH, and MDIH

Table 2 summarizes the results of the five tour construction heuristics NIH, CIH, LIH, FIH, and MDIH, with MDIH configured to use the fifth initial tour establishment rule. (The results for MDIH are already in Table 1but are copied into Table 2 for convenient comparison.) From the results it can be seen that performance of MDIH is best among all heuristics on all five criteria. For example, the average deviation from the optimal solution for MDIH is 4.64%, and there is a relatively sharp decline until we come to the second best on this category, which is FIH at 7.27%. It seems that the performance of FIH is second best (to MDIH) in all criteria except standard deviation. Meanwhile, LIH seems to show the worst performance among those examined, on all criteria except the average deviation from the best solution. Closer inspection of the full results, however, shows that the relatively poor figures for LIH derive from a notable failure on only one problem instance, BRG180. On this instance its best solution is 260% (average + 22 standard deviations) away from the optimal, and its worst solution is 2032.82% (average + 59 standard deviations) away from the optimal. This dataset brings out the worst performance from each of the heuristics and is clearly a particularly unusual example with a structure that deviates from the vast majority of TSPs. In fact this problem arises from a participant-comparison metric in a cards-playing tournament, and performance of any heuristic on it is therefore not reflective of performance on cases where the matrix contents are more closely associated with physical distances. We therefore decided that omitting this dataset would lead to a more useful cross-comparison of the heuristics, and the outcome of that for the remaining 108 datasets is given in Table 3; MDIH is preserved as the best performer in all categories, while LIH is now third best.

Table 2: Results of 5 different tour construction heuristics

Name	Average of best solution % age	Average of worst solution % age	Average of average solution % age	Average of standard deviation % age	Average of execution time seconds
(0)	(1)	(2)	(3)	(4)	(5)
NIH	17.91	25.74	21.17	2.55	0.62
CIH	14.27	18.90	16.64	1.27	0.48
LIH	11.18	34.48	23.69	5.93	0.42
FIH	7.27	14.96	10.16	1.99	0.63
MDIH	4.64	8.59	6.34	0.99	0.46

Table 3: Results of 5 different tour construction heuristics after removing noise (dataset BRG180)

Name	Average of best solution % age	Average of worst solution % age	Average of average solution % age	Average of standard deviation % age	Average of execution time seconds
(0)	(1)	(2)	(3)	(4)	(5)
NIH	17.78	22.37	20.10	1.25	0.63
CIH	14.26	18.82	16.61	1.25	0.48
LIH	8.88	15.97	12.01	1.74	0.42
FIH	6.84	12.89	9.51	1.52	0.64
MDIH	4.35	8.18	6.00	0.95	0.46

In Figure 2 we attempt to visualise the performance of these five heuristics as a function of the number of cities, with number of cities on the horizontal axis, the vertical axis being percent difference of average solution of 30 simulations from the optimal solution. Since there is dense accumulation of datasets up to 2000 cities, the picture is not very clear up to this point. However after this point the pattern is quite clear. It can be seen that heuristics follow two different patterns of rise and falls of graph. One pattern is followed by MDIH, FIH, and LIH, while CIH and NIH follow another pattern. The MDIH-FIH-LIH pattern is dominated by MDIH, followed by FIH. While the CIH-NIH pattern is dominated by CIH. Notably, we can see that MDIH is not bettered by any of the other heuristics as we move beyond the 2000 city mark.

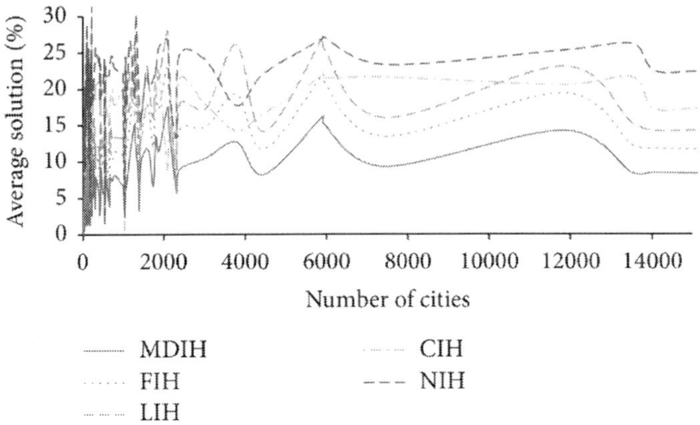

Figure 2: Graph showing average solution performance of heuristics.

To make the picture clear for datasets up to the 2000 cities another graph in Figure 3 is presented. This graph removes the results of datasets more than 2000 cities to make picture clear for smaller problems. However even in this graph picture is not clear up to 200 cities as there is a dense accumulation of datasets up to this point. However picture beyond 200 mark is quiet clear. Again seemingly there are two patterns, one followed by CIH-NIH and another followed by MDIH-FIH-LIH. Former is dominated by CIH and latter is dominated by MDIH. In overall picture mostly MDIH graph dominates all the other graphs.

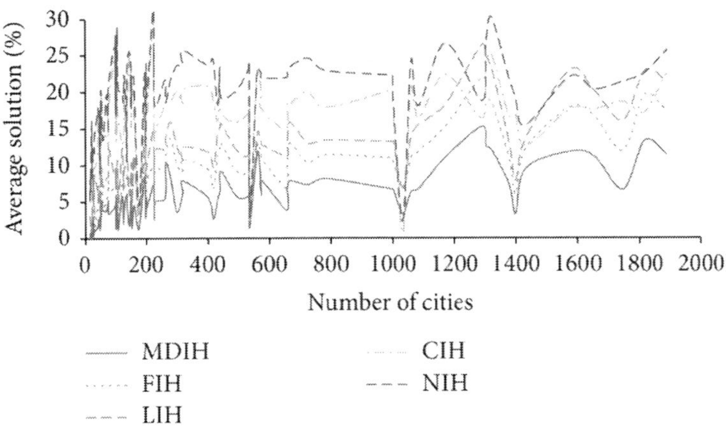

Figure 3: Graph showing average solution performance of heuristics up to 2000 city datasets.

Experiments to Compare Performance of Cost Based Insertion Heuristics with Their Faster Counterparts

We turn now to the fast approximate variants of these heuristics. Table 4 shows comparative results for the standard and fast/approximate variants of CIH, LIH, and MDIH. From the table it can be seen that there are statistically only small differences between the standard and fast variants across all the four criteria. The largest difference between any standard and fast variant occurs in the case of CIH and is in favour of the faster method, with FCIH solution quality on average deviating 13.99% from optimal, compared with 14.26% for CIH. In all cases, the fast method naturally provides a significant advantage in execution time, with a reduction of 70–80% from the execution time of the standard variant.

Table 4: Comparison of standard heuristics with their faster designs

Name	Category	Average of best solution % age	Average of worst solution % age	Average of average solution % age	Average of standard deviation % age	Average of execution time seconds
(0a)	(0b)	(1)	(2)	(3)	(4)	(5)
CIH	Standard	14.26	18.82	16.61	1.25	0.48
	Fast	13.99	18.84	16.46	1.29	0.1
LIH	Standard	8.88	15.97	12.01	1.74	0.42
	Fast	9.02	16.19	12.24	1.79	0.11
MDIH	Standard	4.35	8.18	6.00	0.95	0.46
	Fast	4.3	8.27	6.02	0.99	0.14

We also performed statistical tests (Student's t-test) to investigate whether or not there was any significant difference in solution quality between the standard and faster versions. While comparing results between CIH and FCIH based on a confidence level of 95%, it was found that CIH performed better than FCIH in 17 out of 108 datasets, while FCIH performed better than CIH on 10 out of 108 datasets. Similarly, while comparing results between LIH and FLIH it was found that LIH performed better than FLIH on only 2 out of 108 datasets while FLIH performed better than LIH in 18 out of 108 datasets. While comparing results between MDIH and FMDIH it was found that MDIH performed better than FMDIH in only one out of 108 datasets while FMDIH performed better than MDIH in 3 out of 108 datasets. From these results it is clear that the new heuristics detailed in this paper are competitive in solution quality with the original slower versions; moreover, when there is a statistical difference in quality, this is more often than not (with the exception of CIH versus FCIH) in favor of the faster heuristic. We would conclude that the faster heuristics are generally worth applying to save computational time, with the risk to jeopardise solution quality being balanced by a similar chance of improved solution quality.

We note that a detailed performance comparison among the faster designs is not necessary, since the performance characteristics of the faster designs clearly echo those of the standard designs; therefore comparison among faster versions of heuristics reflects the same comparative analysis that was done among standard versions in Table 3.

Experiments to Compare Performance of FMDIH and FIH

Since MDIH has turned out to be best among all heuristics in the above experiments and its faster version FMDIH has provided same solution quality equal to MDIH with reduced time complexity equal to FIH, this shows a leap forward in the achievement of solution quality with time complexity of (n^2). To test whether or not this advantage is statistically significant we have given comparison of FMDIH with FIH in Appendix with full detail and commentary along with statistical tests. In terms of the columns in Tables 2–4, we simply note here that the FMDIH summary values are, in order of the columns, as follows: 4.30, 8.27, 6.02, 0.99, and 0.14. These are to be compared with the FIH summary values from Table 3, which are, respectively, 6.84, 12.89, 9.51, 1.52, and 0.64. Here however we visualize the relative performance of FMDIH and FIH (previously the best reported (n^2) tour construction heuristic) with Figure 4, again using the same axes as in Figures 2-3. It can be seen in Figure 4 that FMDIH is consistently better than FIH throughout the dataset spectrum, with the advantage even apparent in the dense area of problem instances smaller than 2000 cities.

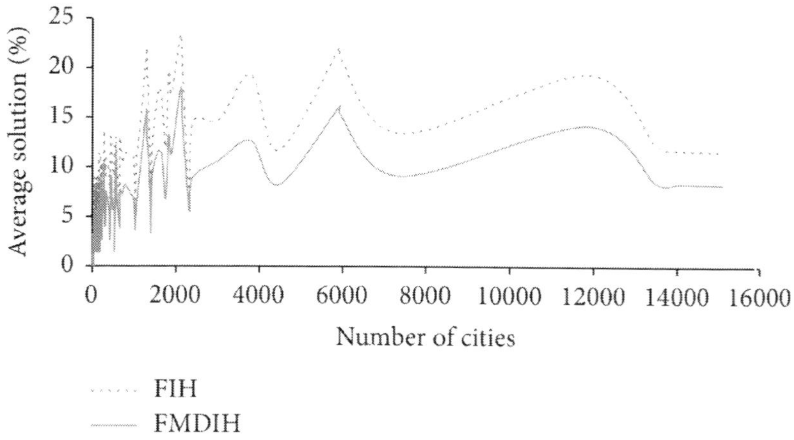

Figure 4: Graph showing performance comparison between FMDIH and FIH.

CONCLUSIONS AND FUTURE WORK

In this paper a previously unpublished tour construction heuristic, namely, the largest insertion heuristic (LIH), is introduced. It is shown that LIH completes the set of well-known popular tour insertion heuristics (cheapest, nearest, and farthest) and seems to produce better solution quality than cheapest and nearest insertion heuristics on the great majority of datasets. Furthermore, this paper also introduces max difference insertion heuristic (MDIH). It is based on a max difference concept previously introduced in [11] as a way to engineer variants of other heuristics; however in this paper we have argued that, following its "completion" by defining an initial subtour establishment method, it merits "first-class" status as a heuristic in itself, alongside others of the same type.The resulting heuristic MDIH produced better solutions on average than all the other baseline heuristics considered in this paper. Furthermore, we have introduced complexity curtailing techniques that reduced worst case complexity of cost based insertion heuristics, that is, CIH, LIH, and MDIH, from (n^3) to $O(n^2$), resulting in the development of their faster versions FCIH, FLIH, and FMDIH. It is shown that the faster versions do not statistically produce inferior solutions but reduce computational cost up to 80%. The development of FMDIH proved a leap forward in production of quality solutions among community of (n^2) complexity heuristics. This is because it has performed 3.5% on average better than FIH on a wide spectrum of popular datasets. We believe this is the first time that FIH has been demonstrated to be outperformed by an (n^2) heuristic on a wide spectrum of popular test bed. Our immediate future work in this direction is inspired by complexity curtailing techniques. What if these techniques are introduced in other higher order heuristics to reduce their complexity. We believe more interesting results are yet to come.

APPENDIX

Comparison between FIH and FMDIH

This appendix presents detailed dataset-wise comparison between two tour construction heuristics, that is, farthest insertion heuristic (FIH) and fast max difference insertion heuristic (FMDIH). Both heuristics have complexity of the order (n^2). However FMDIH designed in this paper has shown consistently better performance in the wide spectrum of datasets in the popular test bed. In Table 5 column-2 shows the results of FIH and column-3 shows the results of FMDIH. Column (a) shows best value among 30 simulations and its percentage difference from the optimal value. Column (b) shows worst value among 30 simulations and its percentage difference from the optimal value. Column (c) shows average value of 30 simulations and its percentage difference from the optimal value. Column (d) shows standard deviation of 30 simulations and its value as a percentage of the optimal value. Column (e) shows average execution time of 30 simulations. The datasets are arranged according to size from smallest (14 cities) to largest (15112 cities). From the column (a) of the table it can be seen that FMDIH has obtained optimal solution in 9 datasets against 7 datasets by FIH. FMDIH has obtained better solution than FIH in 94 out of 108 datasets and in 6 datasets they have solution of same quality. This means that FMDIH has worse solution than FIH in only 8 datasets out of 108 datasets. From the column (b) of the table it can be seen that FMDIH has worse solutions of better quality than FIH in 105 out of 108 datasets. In two datasets they have worse solutions of equal quality. It means only in one dataset FIH has worse solution better than FMDIH. From column (c) of the table it can be seen that average solution of FMDIH is better in 105 out of 108 datasets. This means FMDIH is beaten only in 3 datasets in terms of average solution. From the column

(d) it can be seen that FMDIH has better standard deviation than FIH in 96 out of 108 datasets. This means only on 12 datasets FMDIH is worse in standard deviation than FIH. The story does not stop here; there are more interesting comparisons. By comparing column 3C with column 2a it can be seen that FMDIH has obtained average solutions in 69 out of 108 datasets that are better than the best solutions of FIH. However FIH has not attained this feet on any of the datasets. By comparing column 3b with 2c it can be seen that worst solutions of FMDIH are better than the average solution of FIH in 79 out of 108 datasets, while in none of the datasets FIH has attained this feet. Now most interesting result can be seen by comparing column 3b with column 2a. The worst solutions obtained by FMDIH in 30 simulations are better than the best solutions obtained by FIH in 30 simulations in 28 out of 108 datasets (>25%). If results are looked at more carefully then it can be seen that these 28 datasets start from RD400 dataset. There are only 49 datasets that are of the size 400 cities and above. In 28 out of these 49 datasets (>50%) worst solutions of FMDIH are better than the best solutions of FIH. If dataset spectrum is squeezed to larger band of 1000 cities and above then it is very surprising to see FMDIH has attained worse solutions better than the best solutions of FIH in 25 out of 31 datasets (>80%). Finally in all the 8 datasets of greater than 4000 cities worst solutions of FMDIH are better than the best solutions of FIH in 30 simulations. These results show that advantage to FMDIH improves with increase in size of the dataset and this clearly put performance of FMDIH much ahead of FIH. Student's t-test was conducted which confirmed that the output of FMDIH is better than FIH up to the 99.5% of confidence level on the 100 datasets. On another dataset FMDIH had better results up to 95% of confidence level, where FIH did not prove any statistical supremacy on any of the 108 datasets.

Table 5

Dataset	Optimal	Farthest insertion heuristic									Fast max difference insertion heuristic								
(1)		(2)									(3)								
Min		% diff.	Max	% diff.	Avg.	% diff.	Stdv	% age	Time	Min	% diff.	Max	% diff.	Avg.	% diff.	Stdv	% age	Time	
(a)	(b)	(a)		(b)		(c)		(d)		(e)	(a)		(b)		(c)		(d)		(e)
Burma14	3323	3323	0.00	3442	3.58	3327.83	0.15	21.45	0.65	0.00	3323	0.00	3442	3.58	3326.97	0.12	21.36	0.64	0.00
Ulysses16	6859	6859	0.00	7082	3.25	6900.67	0.61	59.97	0.87	0.00	6859	0.00	7082	3.25	6912.50	0.78	58.89	0.86	0.00
Gr17	2085	2095	0.48	2162	3.69	2102.17	0.82	19.96	0.96	0.00	2085	0.00	2159	3.55	2095.73	0.51	21.58	1.04	0.00
GR21	2707	2707	0.00	2816	4.03	2754.80	1.77	48.57	1.79	0.00	2707	0.00	2707	0.00	2707.00	0.00	0.00	0.00	0.00
Ulysses22	7013	7013	0.00	7592	8.26	7116.03	1.47	131.65	1.88	0.00	7013	0.00	7262	3.55	7088.67	1.08	62.04	0.88	0.00
GR24	1272	1272	0.00	1387	9.04	1322.97	4.01	26.50	2.08	0.00	1272	0.00	1341	5.42	1313.93	3.30	15.82	1.24	0.00
Fri26	937	937	0.00	1021	8.96	957.77	2.22	20.49	2.19	0.00	937	0.00	958	2.24	941.40	0.47	8.00	0.85	0.00
Bayg29	1610	1628	1.12	1746	8.45	1664.53	3.39	28.17	1.75	0.00	1618	0.50	1668	3.60	1631.97	1.36	10.48	0.65	0.00
Bays29	2020	2028	0.40	2083	3.12	2035.80	0.78	12.44	0.62	0.00	2033	0.64	2068	2.38	2038.40	0.91	6.08	0.30	0.00
Dantzig42	699	712	1.86	761	8.87	738.83	5.70	13.07	1.87	0.00	705	0.86	743	6.29	711.63	1.81	8.70	1.25	0.00
Swiss42	1273	1319	3.61	1439	13.04	1371.40	7.73	31.35	2.46	0.00	1273	0.00	1346	5.73	1302.67	2.33	25.60	2.01	0.00
ATT48	10628	10845	2.04	11554	8.71	11205.10	5.43	193.80	1.82	0.00	10653	0.24	11073	4.19	10929.70	2.84	123.93	1.17	0.00
GR48	5046	5099	1.05	5441	7.83	5271.83	4.48	114.17	2.26	0.00	5109	1.25	5295	4.93	5195.73	2.97	61.40	1.22	0.00
HK48	11461	11474	0.11	12513	9.18	11998.70	4.69	316.11	2.76	0.00	11556	0.83	11982	4.55	11611.10	1.31	94.18	0.82	0.00
Eil51	426	437	2.58	462	8.45	450.77	5.81	6.99	1.64	0.00	434	1.88	443	3.99	439.67	3.21	2.33	0.55	0.00
Berlin52	7542	7542	0.00	8677	15.05	8214.50	8.92	316.31	4.19	0.00	7604	0.82	8425	11.71	7948.97	5.40	239.48	3.18	0.00
Brazil58	25395	25664	1.06	27452	8.10	26529.20	4.47	378.63	1.49	0.00	25717	1.27	26546	4.53	26298.90	3.56	196.66	0.77	0.00
ST70	675	696	3.11	736	9.04	710.30	5.23	9.85	1.46	0.00	682	1.04	717	6.22	700.30	3.75	9.18	1.36	0.00
Eil76	538	570	5.95	598	11.15	580.40	7.88	9.64	1.79	0.00	556	3.35	577	7.25	566.20	5.24	5.40	1.00	0.00
PR76	108159	110516	2.18	126235	16.71	115154.00	6.47	4255.54	3.93	0.00	109796	1.51	115392	6.69	111760.00	3.33	1756.05	1.62	0.00
GR96	55209	56854	2.98	63009	14.13	59177.70	7.19	1419.23	2.57	0.00	55738	0.96	59941	8.57	57658.00	4.44	801.67	1.45	0.00

Rat99	1211	1257	3.80	1382	14.12	1313.20	8.44	32.08	2.65	0.00	1244	2.73	1296	7.02	1272.03	5.04	12.63	1.04	0.00
KroA100	21282	22026	3.50	23487	10.36	22608.10	6.23	392.1[illegible]	1.84	0.00	21478	0.92	22599	6.19	21628.70	1.63	217.73	1.02	0.00
KroB100	22141	22693	2.49	24518	10.74	23508.10	6.17	396.25	1.79	0.00	22211	0.32	23405	5.71	22949.90	3.65	282.76	1.28	0.00
KroC100	20749	21105	1.72	22599	8.92	21853.70	5.32	375.72	1.81	0.00	20749	0.00	21838	5.25	21045.70	1.43	225.82	1.09	0.00
KroD100	21294	21745	2.12	23374	9.77	22528.60	5.80	376.69	1.77	0.00	21403	0.51	22708	6.64	21880.20	2.75	286.35	1.34	0.00
KroE100	22068	22514	2.02	24147	9.42	23229.20	5.26	484.60	2.20	0.00	22179	0.50	23195	5.11	22749.20	3.09	300.23	1.36	0.00
RD100	7910	8255	4.36	8936	12.97	8619.23	8.97	167.00	2.11	0.00	7965	0.70	8380	5.94	8158.47	3.14	111.72	1.41	0.00
Eil101	629	660	4.93	701	11.45	679.60	8.04	10.61	1.69	0.00	648	3.02	683	8.59	664.60	5.66	7.90	1.26	0.00
Lin105	14379	14863	3.37	16088	11.89	15435.30	7.35	309.15	2.15	0.00	14402	0.16	15179	5.56	14708.30	2.29	182.60	1.27	0.00
PR107	44303	45030	1.64	45933	3.68	45450.50	2.59	245.78	0.55	0.00	44489	0.42	45420	2.52	44925.80	1.41	216.48	0.49	0.00
GR120	6942	7162	3.17	7817	12.60	7462.97	7.50	162.80	2.35	0.00	7164	3.20	7494	7.95	7323.47	5.50	79.69	1.15	0.00
PR124	59030	60305	2.16	65845	11.54	62875.70	6.51	1339.81	2.27	0.00	59767	1.25	62202	5.37	60323.20	2.19	625.42	1.06	0.00
Bier127	118282	123796	4.66	130733	10.53	126795.00	7.20	1891.57	1.60	0.00	121130	2.41	131232	10.95	123455.00	4.37	1996.69	1.69	0.00
CH130	6110	6335	3.68	6884	12.67	6556.80	7.31	139.36	2.28	0.00	6328	3.57	6686	9.43	6441.50	5.43	78.33	1.28	0.00
PR136	96772	102573	5.99	107287	10.87	105143.00	8.65	1129.12	1.17	0.00	99960	3.29	104166	7.64	102520.00	5.94	975.59	1.01	0.00
GR137	69853	73391	5.06	81011	15.97	76135.90	8.99	1606.32	2.30	0.00	71773	2.75	74864	7.17	72760.50	4.16	785.68	1.12	0.00
PR144	58537	59336	1.36	64705	10.54	62041.30	5.99	1383.72	2.36	0.00	59005	0.80	63056	7.72	59680.10	1.95	807.68	1.38	0.00
CH150	6528	6864	5.15	7479	14.57	7094.67	8.68	148.20	2.27	0.00	6733	3.14	7103	8.81	6901.07	5.71	97.47	1.49	0.00
kroA150	26524	27407	3.33	29394	10.82	28514.90	7.51	544.26	2.05	0.00	27066	2.04	28285	6.64	27593.20	4.03	284.44	1.07	0.00
kroB150	26130	27337	4.62	28912	10.65	27859.30	6.62	375.99	1.44	0.00	26322	0.73	27868	6.65	27049.60	3.52	392.87	1.50	0.00
PR152	73682	74827	1.55	77967	5.82	76401.20	3.69	808.26	1.10	0.00	74029	0.47	76257	3.49	74715.40	1.40	615.89	0.84	0.00
U159	42080	44506	5.77	49335	17.24	46856.30	11.35	1195.20	2.84	0.00	43367	3.06	45621	8.41	44070.50	4.73	527.23	1.25	0.00
SI175	21407	21751	1.61	22193	3.67	21911.90	2.36	119.10	0.56	0.00	21527	0.56	21883	2.22	21707.30	1.40	112.23	0.52	0.00
Rat195	2323	2503	7.75	2686	15.63	2585.10	11.28	42.01	1.81	0.00	2487	7.06	2588	11.41	2525.47	8.72	25.35	1.09	0.00
d198	15780	16277	3.15	17207	9.04	16696.60	5.81	311.97	1.98	0.00	16015	1.49	16621	5.33	16214.40	2.75	161.35	1.02	0.00
KroA200	29368	30533	3.97	32481	10.60	31225.90	6.33	448.47	1.53	0.00	30258	3.03	31662	7.81	31002.70	5.57	372.36	1.27	0.00
KroB200	29437	31227	6.08	33304	13.14	32006.80	8.73	427.13	1.45	0.00	30387	3.23	31877	8.29	30987.90	5.27	438.01	1.49	0.00
GR202	40160	42038	4.68	45263	12.71	43423.10	8.13	725.72	1.81	0.00	40889	1.82	43698	8.81	41942.20	4.44	646.74	1.61	0.00

TS225	126643	134365	6.10	143384	13.22	139859.00	10.44	2128.75	1.68	0.00	135470	6.97	138995	9.75	136429.00	7.73	1115.29	0.88	0.00
TSP225	3916	4182	6.79	4501	14.94	4292.40	9.61	78.76	2.01	0.00	4118	5.16	4292	9.60	4187.50	6.93	45.32	1.16	0.00
PR226	80369	81020	0.81	84593	5.26	82441.70	2.58	848.57	1.06	0.00	81598	1.53	83216	3.54	82492.30	2.64	296.77	0.37	0.00
GR229	134602	142714	6.03	154567	14.83	147270.00	9.41	3066.84	2.28	0.00	138704	3.05	145318	7.96	141429.00	5.07	1740.97	1.29	0.00
Gil262	2378	2507	5.42	2683	12.83	2597.10	9.21	42.29	1.78	0.00	2448	2.94	2548	7.15	2514.13	5.72	24.39	1.03	0.00
PR264	49135	53814	9.52	55504	12.96	54821.70	11.57	425.79	0.87	0.00	53021	7.91	55170	12.28	54211.90	10.33	508.62	1.04	0.00
A280	2579	2847	10.39	3043	17.99	2931.93	13.68	50.88	1.97	0.00	2691	4.34	2861	10.93	2792.60	8.28	44.26	1.72	0.00
PR299	48191	51625	7.13	56160	16.54	53082.40	10.15	985.20	2.04	0.00	48763	1.19	52054	8.02	50174.00	4.11	747.97	1.55	0.00
Lin318	42029	44730	6.43	46981	11.78	45840.00	9.07	575.64	1.37	0.00	43516	3.54	45306	7.80	44471.20	5.81	506.01	1.20	0.00
Linhp318	41345	44730	8.19	46981	13.63	45840.00	10.87	575.64	1.39	0.00	43516	5.25	45306	9.58	44471.20	7.56	506.01	1.22	0.00
RD400	15281	16441	7.59	17139	12.16	16724.50	9.45	164.65	1.08	0.00	15876	3.89	16431	7.53	16132.40	5.57	128.14	0.84	0.00
FL417	11861	12201	2.87	13147	10.84	12657.10	6.71	245.19	2.07	0.00	12061	1.69	12214	2.98	12171.80	2.62	32.24	0.27	0.00
GR431	171414	181584	5.93	200082	16.72	191434.00	11.68	5458.44	3.18	0.00	177362	3.47	183560	7.09	180437.00	5.26	1412.80	0.82	0.00
PR439	107217	116581	8.73	124415	16.04	120425.00	12.32	2176.87	2.03	0.00	110900	3.44	117753	9.83	114517.00	6.81	1734.12	1.62	0.00
PCB442	50778	55660	9.61	59032	16.26	57419.30	13.08	736.18	1.45	0.00	54452	7.24	56840	11.94	55421.30	9.14	550.44	1.08	0.00
D493	35002	37545	7.27	39044	11.55	38212.00	9.17	440.14	1.26	0.00	36402	4.00	37732	7.80	36984.00	5.66	294.40	0.84	0.00
ATT532	27686	29670	7.17	31101	12.33	30257.50	9.29	360.98	1.30	0.00	28797	4.01	29796	7.62	29293.20	5.81	242.24	0.87	0.00
Ali535	202339	216002	6.75	238252	17.75	227611.00	12.49	6102.76	3.02	0.01	210768	4.17	221967	9.70	216135.00	6.82	2858.38	1.41	0.01
SI535	48450	49496	2.16	50025	3.25	49816.20	2.82	148.87	0.31	0.00	49020	1.18	49346	1.85	49167.90	1.48	83.64	0.17	0.00
PA561	2763	3061	10.79	3204	15.96	3118.40	12.86	29.61	1.07	0.01	3007	8.83	3141	13.68	3092.17	11.91	33.81	1.22	0.01
U574	36905	40097	8.65	42081	14.03	40969.00	11.01	370.18	1.00	0.01	38752	5.00	39928	8.19	39252.00	6.36	346.62	0.94	0.01
Rat575	6773	7374	8.87	7594	12.12	7508.17	10.85	53.59	0.79	0.01	7191	6.17	7444	9.91	7300.83	7.79	58.16	0.86	0.01
P654	34643	35659	2.93	41131	18.73	37016.60	6.85	1376.23	3.97	0.01	35334	1.99	36176	4.43	35971.10	3.83	207.24	0.60	0.01
D657	48912	52482	7.30	55079	12.61	53826.00	10.05	650.07	1.33	0.01	51459	5.21	53624	9.63	52371.20	7.07	430.21	0.88	0.01
GR666	294358	324349	10.19	341937	16.16	332601.00	12.99	4859.02	1.65	0.01	311264	5.74	328227	11.51	317068.00	7.72	3965.46	1.35	0.01
U724	41910	45425	8.39	47487	13.31	46237.80	10.33	470.09	1.12	0.01	44499	6.18	45481	8.52	44946.70	7.25	277.31	0.66	0.01
Rat783	8806	9607	9.10	9959	13.09	9809.83	11.40	70.02	0.80	0.01	9419	6.96	9665	9.75	9532.33	8.25	62.85	0.71	0.01

DSJ1000	18659688	2.04E + 07	9.33	2.12E + 07	13.46	20705300.00	10.96	199033.00	1.07	0.03	1.97E + 07	5.63	2.02E + 07	8.51	19916900.00	6.74	138698.00	0.74	0.02
DSJ1000_ Ceil	18660188	2.04E + 07	9.33	2.12E + 07	13.46	20705900.00	10.96	199248.00	1.07	0.03	1.97E + 07	5.63	2.02E + 07	8.51	19914600.00	6.72	136854.00	0.73	0.02
Pr1002	259045	279938	8.07	290550	12.16	285664.00	10.28	2895.93	1.12	0.03	270964	4.60	280759	8.38	276781.00	6.85	2215.42	0.86	0.02
Si1032	92650	98444	6.25	99088	6.95	98819.50	6.66	154.80	0.17	0.03	95552	3.13	96430	4.08	96050.60	3.67	196.73	0.21	0.02
U1060	224094	245679	9.63	254238	13.45	248951.00	11.09	2035.57	0.91	0.03	235078	4.90	241873	7.93	238367.00	6.37	1861.23	0.83	0.02
Vm1084	239297	262043	9.51	274616	14.76	267609.00	11.83	2850.55	1.19	0.03	253819	6.07	262898	9.86	256516.00	7.20	1871.48	0.78	0.02
Pcb1173	56892	64918	14.11	66795	17.41	65640.20	15.38	442.92	0.78	0.04	62530	9.91	64149	12.76	63103.30	10.92	358.86	0.63	0.02
D1291	50801	60221	18.54	63234	24.47	61956.00	21.96	614.26	1.21	0.05	57501	13.19	59945	18.00	58810.80	15.77	641.59	1.26	0.03
RL1304	252948	292602	15.68	308095	21.80	301514.00	19.20	3688.25	1.46	0.05	279047	10.32	291228	15.13	285207.00	12.75	3363.99	1.33	0.03
RL1323	270199	312400	15.62	328087	21.42	320036.00	18.44	4117.85	1.52	0.05	297661	10.16	307528	13.82	302587.00	11.99	2445.58	0.91	0.03
NRW1379	56638	62208	9.83	63899	12.82	62863.50	10.99	420.18	0.74	0.06	60170	6.24	61393	8.40	60785.70	7.32	279.13	0.49	0.03
FL1400	20127	20949	4.08	21909	8.85	21338.30	6.02	273.56	1.36	0.06	20516	1.93	21201	5.34	20786.70	3.28	213.61	1.06	0.03
U1432	152970	168810	10.35	174677	14.19	171655.00	12.21	1166.59	0.76	0.06	165373	8.11	169884	11.06	167195.00	9.30	1092.88	0.71	0.03
FL1577	22249	25155	13.06	27824	25.06	26228.30	17.89	514.47	2.31	0.07	24030	8.00	25892	16.37	24835.70	11.63	445.83	2.00	0.04
D1655	62128	71507	15.10	73675	18.59	72508.90	16.71	530.78	0.85	0.08	68217	9.80	70258	13.09	69165.90	11.33	447.10	0.72	0.04
VM1748	336556	371239	10.31	381960	13.49	376862.00	11.98	3094.47	0.92	0.10	355083	5.50	364325	8.25	359501.00	6.82	2503.97	0.74	0.05
U1817	57201	67612	18.20	69628	21.73	68406.70	19.59	596.43	1.04	0.10	63383	10.81	65642	14.76	64709.40	13.13	519.68	0.91	0.05
RL1889	316536	365077	15.34	376387	18.91	371580.00	17.39	3230.94	1.02	0.12	348074	9.96	359023	13.42	352505.00	11.36	2975.08	0.94	0.06
D2103	80450	98109	21.95	100527	24.96	99264.90	23.39	657.60	0.82	0.14	93544	16.28	95910	19.22	94908.90	17.97	527.46	0.66	0.07
U2152	64253	75721	17.85	77597	20.77	76732.50	19.42	527.56	0.82	0.15	72492	12.82	74131	15.37	73311.70	14.10	430.31	0.67	0.07
U2319	234256	247881	5.82	251176	7.22	250028.00	6.73	715.57	0.31	0.17	245939	4.99	248584	6.12	247465.00	5.64	687.33	0.29	0.09
Pr2392	378032	428627	13.38	438782	16.07	432695.00	14.46	2243.77	0.59	0.19	406210	7.45	417522	10.45	411749.00	8.92	2681.97	0.71	0.09
Pcb3038	137694	156708	13.81	159518	15.85	158069.00	14.80	586.58	0.43	0.32	151346	9.91	153766	11.67	152435.00	10.71	603.48	0.44	0.15
FL3795	28772	33124	15.13	35354	22.88	34331.30	19.32	542.79	1.89	0.49	31317	8.85	33137	15.17	32407.00	12.63	585.13	2.03	0.24
FNL4461	182566	202328	10.82	204937	12.25	203925.00	11.70	601.55	0.33	0.74	196918	7.86	198936	8.97	197617.00	8.24	495.01	0.27	0.32
RL5915	565530	683054	20.78	697158	23.28	690217.00	22.05	3493.67	0.62	1.39	648152	14.61	662852	17.21	656416.00	16.07	3429.91	0.61	0.56
RL5934	556045	660399	18.77	681665	22.59	672930.00	21.02	4608.88	0.83	1.40	630512	13.39	649307	16.77	641010.00	15.28	3805.45	0.68	0.56

pla7397	23260728	26066700	12.06	26640000	14.53	26382900.00	13.42	143431.00	0.62	2.32	25178800	8.25	25559100	9.88	25383600.00	9.13	87578.50	0.38	0.86
rl11849	923288	1093120	18.39	1108780	20.09	1102520.00	19.41	3886.70	0.42	10.11	1046110	13.30	1060200	14.83	1054670.00	14.23	3259.40	0.35	2.23
usa13509	19982859	22318200	11.69	22561500	12.90	22442600.00	12.31	59189.90	0.30	15.01	21566100	7.92	21768200	8.93	21673900.00	8.46	50429.90	0.25	2.88
brd14051	469385	522027	11.22	527320	12.34	524714.00	11.79	1160.75	0.25	16.43	506838	7.98	510569	8.77	508872.00	8.41	910.74	0.19	3.12
d15112	1573084	1750830	11.30	1762650	12.05	1756490.00	11.66	2965.69	0.19	19.11	1698080	7.95	1707830	8.57	1704060.00	8.33	2482.61	0.16	3.61
			6.84		12.89		9.51		1.52	0.64		4.30		8.27		6.02		0.99	0.14

ACKNOWLEDGMENTS

The authors are grateful for financial support from Innovate UK and Route Monkey Ltd. via KTP partnership no. 9839.

REFERENCES

1. D. L. Applegate, R. E. Bixby, V. Chvátal, and W. J. Cook, The Traveling Salesman Problem, Princeton University Press, 2007.
2. J. K. Lenstra and A. H. G. R. Kan, "Some simple applications of the travelling salesman problem," Operational Research Quarterly, vol. 26, no. 4, pp. 717–733, 1975. ·
3. S. Lin and B. W. Kernighan, "An effective heuristic algorithm for the traveling-salesman problem," Operations Research, vol. 21, no. 2, pp. 498–516, 1973. · · ·
4. http://www.math.uwaterloo.ca/tsp/concorde/index.html.
5. J. Perttunen, "On the significance of the initial solution in travelling salesman heuristics," The Journal of the Operational Research Society, vol. 45, no. 10, pp. 1131–1140, 1994.
6. D. J. Rosenkrantz, R. E. Stearns, and I. Lewis, "An analysis of several heuristics for the traveling salesman problem," SIAM Journal on Computing, vol. 6, no. 3, pp. 563–581, 1977. · · ·
7. D. S. Johnson and L. A. McGeoch, "Experimental analysis of heuristics for the STSP," in The Traveling Salesman Problem and Its Variations, G. Gutin and A. P. Punnen, Eds., vol. 12 of Combinatorial Optimization, chapter 9, pp. 369–443, Kulwer Academic Publishers, 2002. · ·
8. J. J. Bently, "Fast algorithms for the geometric travelling salesman problem," ORSA Journal on Computing, vol. 4, no. 4, pp. 387–411, 1992. ·
9. D. S. Johnson and L. A. McGeoch, "The traveling salesman problem: a case study in local optimization," in Local Search in Combinatorial Optimization, vol. 1, pp. 215–310, John Wiley & Sons, 1997.
10. T. A. J. Nicholson, "A sequential method for discrete optimization problems and its application to the assignment, travelling salesman, and three machine scheduling problems," IMA Journal of Applied

Mathematics, vol. 3, no. 4, pp. 362–375, 1967. · ·

11. G. Righini, The Largest Insertion algorithm for the Traveling Salesman Problem, 2000, http://crema.di.unimi.it/~righini/Didattica/Algoritmi%20Euristici/MaterialeAE/larg_ins.pdf.
12. T. W. Tunnel and L. Heath, Development of new heuristics for the Euclidean travelling salesman problem [M.S. thesis], Virginia Polytechnic Institute and State University, Blacksburg, Va, USA, 1989, http://eprints.cs.vt.edu/archive/00000167/01/TR-89-30.pdf.
13. G. Reinelt, The Traveling Salesman Problem: Computational Solutions for TSP Applications, vol. 840 of Lecture Notes in Computer Science, Springer, Berlin, Germany, 1994.
14. B. L. Golden and W. Stewart Jr., "Empirical analysis of heuristics," in The Travelling Salesman Problem, E. L. Lawler, J. K. Lenstra, A. H. G. Rinnoy Kan, and D. B. Shimoys, Eds., John Wiley & Sons, London, UK, 1985.

REFERENCES

1. Abbass, H.A., Sarker, R., and Newton, C., "PDE: a Pareto-frontier differential Evolution approach formulti-objective optimization problems, "presented at Evolutionary Computation, 2001. Proceedings of the 2001 Congress on, Seoul, South Korea, 2001.
2. AGUIAR, F., CREATED, H. And HERREROS, F. (2003), "Sociology and rational choice" in S. Giner (comp.). Modern sociological theory. Ariel.
3. ALLAIS, M. and HAGEN, J. (eds., 1979), Expected Utility Hypotheses and the Allais Paradox. Dordrecht: Reidel
4. ANAND, P. (1993), Foundations of Rational Cholee Under Risk. Oxfrod: Clarendon Press.
5. BARBERA, S. (1984), "Theory of social choice: some lines of development", Scienda Spanish Public, 91: 221-243.
6. BAUMOL, W.J. (1911), Economic Theory and Operation Analysis. London: Prentice- Hall.
7. Benayoun, R., Montgolfier, J., Tergny, J., and Laritchev, O., "Linear

Programming with Multiple Objective Functions: Step Method (STEM), “Mathematical Programming, Vol. 1, pp. 366-375, 1971.

8. BECKER, G. (1986) “The economic approach to human behavior”, in Elster, J. (comp.), Rational Cholee. Oxford: Basil Blackwell, pp. 108-122.
9. BELL, D., RAIFFA, H. and TVERSKY, A. (1988) “Descriptive, normative and prescriptive Interactions in decision making “, in Bell, Raiffa and Tversky (comps.). Decision Making. Cambridge: Cambridge University Press, pp. 9-32.
10. BELL, D., RAIFFA, H. and TVERSKY, A. (1988), Decision Making. Cambridge: Cambridge University Press.
11. BINMORE, K. (1994), Game Theory. Madrid: McGraw-Hill.
12. BRODER, A. (2000), “Assesing the empirical validity of” Take The Best “heuristic as a Model of human probabilistic inference ‘, Journal of Experimental Psychology: Learning, Merrill, and Cognition, 26: 1332-1346.
13. CAÑAS, J. and HARO, T. (1982), “The management of stocks in a context of uncertainty” Journal of Political Economy, 90: 179-200.
14. Caballero, R., Luque, M., Molina, J., and Ruiz, F., “Multiobjective Programming Interactiva, “Vol. Monograph, No. 1, pp. 123 -148, 2002.
15. Charnes, A., Cooper, W., and Ferguson, R., “Optimal estimation of executive Compensation by linear programming, “Management Science, vol.1, pp. 138-151, 1955.
16. Coelho, R.F., Bersini, H., and Bouillard, P., “Parametrical mechanical design with Constraints and preferences: application to a purge valve, “Computer methods in Applied mechanics and engineering, vol. 192, pp. 4355-4378, 2003.
17. Coello, C. C., “An Updated Survey of Evolutionary Multiobjective Optimization Techniques: State of the Art and Future Trends, “IEEE, pp. 13, 1999.
18. Coello, C. C., “Handling Preferences in Evolutionary Multiobjective Optimization: A Survey, “presented at Congress on Evolutionary Computation, 2000.
19. Coello, C. C., “Evolutionary Multi-Objective Optimization: A

Critical Review," Evolutionary Optimization, 2002.

20. Coello, C. C., "Twenty Years of Evolutionary Multi-Objective Optimization: A Historical View of the Field, "pp. 20 p., 2005.
21. Coello, C. C., "Evolutionary Multi-Objective Optimization: Past, Present and Future," Presented at Conference at the National University of Colombia, Medellín, 2007.
22. Coello, C.C., Veldhuizen, D.A. V., and Lamont, G. B., Evolutionary Algorithms for Solving Multi-objective Problems. New York: Plenum Publishers, 2002.
23. Coello, C. C., "Introduction to Evolutionary Computing: Course Notes." Mexico: Department of Electrical Engineering, Computer Section. Av. Instituto National Polytechnic, 2004, pp. 310.
24. Cohon, J. L. and Marks, D. H., "A Review and Evaluation of Multiobjective Programming Techniques, "Water Resources Research, vol. 11 (2), pp. 208-220,1975.
25. COLEMAN, J. (1990), Foundations of Social Theory. Cambridge, Mass .: Harvard University Press.
26. COOK, K. and LEVI, M., (eds., 1990), The Limits of Rationality. Chicago: University of Chicago Press.
27. Cvetkoviç, D. and Coello, C.A., "Human Preferences and their Applications in Evolutionary Multi-Objective Optimization, "in RCS version 1.21, Canada, 2004, pp. 24.
28. Deb, K., Multi-Objective Optimization using Evolutionary Algorithms, Second Edition ed. Chichester, UK: John Wiley & Sons, Inc., 2001.
29. Deb, K. and Goel, T., "Controlled Elitist Non-dominated Sorting Genetic Algorithms for Better Convergence, "in Evolutionary Multi-Criterion Optimization, Vol. 1993/2001, Heidelberg, S.B., Ed., 2001, pp. 67-81.
30. Deb, K. and Chaudhuri, S., "we-EMO: An Interactive Evolutionary Multi-Objective Optimization Tool, "in Pattern Recognition and Machine Intelligence, Heidelberg, S. B., Ed. Kanpur, India: Kanpur Genetic Algorithms Laboratory (KanGAL), 2005, pp. 690-695.
31. Deb, K., Pratap, A., Agarwal, S., and Meyarivan, T., "A Fast and Elitist Multiobjective Genetic Algorithm: NSGA-II, "IEEE

Transactions on Evolutionary Computation, vol. 6, pp. 182-197, 2002.

32. DeJong, K.A., "Analysis of the Behavior of a Class of Genetic Adaptive Systems," in Computer and Communication Sciences Department. Michigan: University of Michigan, 1975, pp. 256.
33. Fogel, L.J., Owens, A.J., and Walsh, M.J., "Artificial Intelligence through Simulated Evolution, "John Wiley & Sons, we., Ed. New York, NY, USA, 1966.
34. Fonseca, C.M. and Fleming, P.J., "Genetic algorithms for multiobjective optimization: Formulation, discussion and generalization, "in Genetic Algorithms: Proceedings of The Fifth International Conference, S. Forrest, E., Ed. San Mateo: Morgan Kaufmann, 1993, pp. 416-423.
35. FiSHBURN, P.C. (1988), Nonlinear Preference and Utility Theory. Baltimore: John Hopkins University Press.
36. FRENCH, S. (1986), Decision Theory: An Introduction to the Mathematics of Rationality. Chichester: EUis Horwood Ltd.
37. GIGERENZER, G., Todd, P. et al (2000), Simple Heuristics that Make Us Smart. Oxford: Oxford University Press.
38. GIGERENZER, G. and Selten, R. (2002), Bounded Rationality. The Adaptive Toolbox. Cambridge, Mass: The MIT Press.
39. GOLDSTEIN, WiLLAM M. and HoGARTH, ROBIN M. (eds.) (1997), Research on Judgment And Decision Making: Currents, Connections, and Controversies. Cambridge: Cambridge University Press.
40. HARSANYI, J. (1975). «Can the maximin begin to serve as a basis for moraiity?» American Political Science Review 69 (2): 599-606.
41. HARSANYI, J. (1976). Essays in Ethics, Social Behavior, and Scientific Explanation. Dordreclit, Reidei.
42. HOGARTH, R. M. (1990), Insights in Decision Making. Ciiicago: University of Chicago Press.
43. Hillier, F. and Lieberman, G. J., Operations Research, Seventh ed: McGraw Hill, 2002.
44. Hillier, M. S. and Hillier, F. S., "Conventional Optimization Techniques," in Evolutionary Optimization, Sarker, R.,

Mohammadian, M., and Yao, X., Eds. California: Kluwer Academic Publishers, 2002, pp. 418.

45. Holland, J. H., "Concerning efficient adaptive systems," in Self-Organizing Systems - 1962, M. C. Yovits, Jacobi, G.T., and Goldstein., G.D., Eds. Washington D.C .: Spartan Books, 1962, pp. 215-230.
46. Holland, J. H., "Adaptation in Natural and Artificial Systems," University of Michigan Press, Michigan 1975.
47. Horn, J., Nafpliotis, N., and Goldberg, D.E., "A niched Pareto genetic algorithm for Multiobjective optimization, "Evolutionary Computation, 1994. IEEE World Congress on Computational Intelligence, Proceedings of the First IEEE Conference Vol. 1, pp. 82-87, 1994.
48. Ishibuchi, H., Nojima, Y., Narukawa, K., and Doi, T., "Incorporation of decision makers Preference into evolutionary multiobjective optimization algorithms. " At Genetic and Evolutionary Computation Conference (GECCO'2006), Washington, USA, 2006.
49. Jaramillo, G. P., "Development of a decision support system for allocation Natural resources with the satisfaction of multiple objectives and multiple Makers. "Spain: Polytechnic University of Valencia, 1999.
50. Jin, Y. and Sendhoff, B., "Incorporation of Fuzzy Preferences into Evolutionary Multiobjective Optimization. ", Presented at Proceedings of the 4th Asia Pacific Conference on Simulated Evolution and Learning, Singapore, 2002.
51. Korhonen, P. and Laakso, J., "The Visual Interactive Method for Solving the Multiple Criteria Problem (VIA), "European Journal of Operational Research, vol. 24, pp. 277-287, 1986.
52. KAHNEMAN, D .; Slovic, P. and Tversky, A. (eds., 1982), Judgment under Uncertainty: Heuristics, Cambridge: Cambridge University Press.
53. LUCE, R. and RAIFFA, H. (1957), Games and Decisions. Chichester: Wiley.
54. MARÍ-KLOSE, P. (2000), Rational Election. Madrid: CIS
55. Morrow, J. (1994), Ghent Theory for Political Scientist. Princeton:

Princeton University Press.

56. Munier, B.R. (comp., 1988), Risk, Decision and Rationality. Dordrecht: Reidei
57. PAYNE, J .; BETTMAN, J. and JOHSON, E. (1993), The Adaptive Decision Maker. Cambridge: Cambridge University Press
58. PEACOCK, A. (1995), Public choice: a historical perspective. Madrid: Alianza Editorial.
59. RAMSEY, F. (1931), The Foundations of Mathematics and Other Logical Essays. London: Keagan Paul.
60. RAPOPORT (1983), "An overview of decision theory", in Rapoport, A., Mathematical Models in the Social and Behavioral Sciences. London: John Wiley & Sons.
61. RASMUSEN, E. (1989), Games and Information. Oxfrod: Blackewell.
62. RAWLS, J. (1971), A Theory of Judgment. Oxford: Oxford University Press In Fund of Economic Culture]
63. ROBLES, J. M. (2004), "Ecological Rationality: Heuristic and Blurred Limitations" Unpublished book
64. RUBISNTEIN, A. (1998), Modeling Bounded Rationality. Cambridge, Mass .: The MU Press.
65. Rachmawati, L. and Srinivasan, D., "Preference Incorporation in Multi-objective Evolutionary Algorithms: A Survey, "presented at IEEE Congress on Evolutionary Computation, Vancouver, BC, Canada, 2006.
66. Rechenberg, we., "Cybernetic solution path of an experimental problem," Royal Aircraft Establishment Library Translation no. 122, 1965.
67. Santana, L. V. and Coello, C.A., "An Introduction to Evolutionary Computing and Some of its applications in Economics and Finance, "Journal of Methods Quantitative for the economy and the company., Vol. 2, pp. 3-26, 2006.
68. Schaffer, J.D., "Multiple objective optimization with vector-evaluated genetic Algorithms, "presented at 1st International Conference on Genetic Algorithms and Their Applications, Carnegie-Mellon Univ., Pittsburgh, 1985.

69. Smith, R., Mesa, O., Dyner, we., Jaramillo, P., Poveda, G., and Valencia, D., Decisions With Multiple Objectives and Uncertainty, Second Edition Expanded ed. Medellin: National University of Colombia, 2000.

70. Srinivas, N. and Deb, K., "Multiobjective Optimization Using Nondominated Sorting in Genetic Algorithms, "Journal of Evolutionary Computation, Vol. 2. No. 3, pp. 221- 248, 1994.

71. SELTEN, R. (1996), "Descriptive approaches to cooperation", in S. Hart and A. Mas-Colell (Eds.), Cooperation: Game-Theoretic Approaches. Berlin: Springer, pp. 289-328.

72. SIMÓN, H .; Egidi, M. et al. (1992), Economics, Bounded Rationality and the Cognitive Revolution. Vermont: Edward Elgar.

73. SIMÓN, H. (1986). "From substantive to procedural rationality" in F. Hahn and M. Hollis (Comp.) Philosophy and economic theory. Mexico: FCE, pp. 130-171.

74. SIMÓN, H. (1987), "Rationality in psychology and economics", in R. Hogarth and M. Reder (comps.), Rational Cholee. Chicago: University of Chicago Press.

75. SLOTE, M. (1989). Beyond Optimazing, Cambridge, Mass .: Harvard University Press.

76. TODD, P. and GiGERENZER, G. (2003), "Bounding rationality to the vjorld", Journal of Economics Psychology, vol. 24, no. 2, pp. 143-165.

77. VON NEUMANN, J. and MORGENSTERN, O. (1944), Theory of Games and Economic Behavior. Princeton: Princeton University Press.

78. Veldhuizen, D. A. V. and Lamont, G. B., "Multiobjective Evolutionary Algorithms: Analyzing the State-of-the-Art, "Evolutionary Computation, vol.8, pp. 125-147, 2000.

79. Zadeh, L., "Optimality and non-scalar-valued performance criteria," Automatic Control, IEEE Transactions, vol. 8, p. 59-60, 1963.

80. Zeleny, M., "Compromise Programming in Multiple Criteria Decision Making," University of South Carolina Press, Columbia, USA, pp. 262-301, 1973.

81. Zitzler, E. and Thiele, L., "Multiobjective Evolutionary Algorithms: A Comparative Case Study and the Strength Pareto Approach, "IEEE Transactions On Evolutionary Computation, Vol. 3. No. 4, pp. 257-271, 1999.

82. Zitzler, E., Laumanns, M., and Thiele, L., "SPEA2: Improving the Pareto Strength Evolutionary Algorithm, "Swiss Federal Institute of Technology (ETH), Zurich, Switzerland TIK-Report 103, May 2001.

Citations

CHAPTER 7

João Marques, Maria Cunha, Dragan Savić, and Orazio Giustolisi, "Water Network Design Using a Multiobjective Real Options Framework," Journal of Optimization, vol. 2017, Article ID 4373952, 13 pages, 2017. doi:10.1155/2017/4373952.

CHAPTER 8

Siwar Ben Haj Hassine, Mehdi Jemai, and Bouraoui Ouni, "Power and Execution Time Optimization through Hardware Software Partitioning Algorithm for Core Based Embedded System," Journal of Optimization, vol. 2017, Article ID 8624021, 11 pages, 2017. doi:10.1155/2017/8624021.

CHAPTER 9

Amadeo Ascó, "An Analysis of Robustness Approaches for the Airport Baggage Sorting Station Assignment Problem," Journal of Optimization, vol. 2016, Article ID 1213949, 19 pages, 2016. doi:10.1155/2016/1213949. Chapter 10

CHAPTER 10

S. Morteza Mirdehghan and Hassan Rostamzadeh, "Finding the Efficiency Status and Efficient Projection in Multiobjective Linear Fractional Programming: A Linear Programming Technique," Journal of Optimization, vol. 2016, Article ID 9175371, 8 pages, 2016. doi:10.1155/2016/9175371.

CHAPTER 11

Jiyong Oh, Nojun Kwak, Generalized mean for robust principal component analysis, Pattern Recognition, Volume 54, 2016, Pages 116-127, ISSN 0031-3203, http://dx.doi.org/10.1016/j.patcog.2016.01.002.

CHAPTER 12

Ziauddin Ursani and David W. Corne, "Introducing Complexity Curtailing Techniques for the Tour Construction Heuristics for the Travelling Salesperson Problem," Journal of Optimization, vol. 2016, Article ID 4786268, 15 pages, 2016. doi:10.1155/2016/4786268.

INDEX

G

H

I

L

M

N

O

P

R

S